Springer
Berlin
Heidelberg
New York
Barcelona
Budapest
Hong Kong
London
Milan
Paris
Santa Clara
Singapure
Tokyo

The German Advisory Council on Global Change

(Wissenschaftlicher Beirat der Bundesregierung Globale Umweltveränderungen)

(Members as at June 1, 1996)

Prof. Friedrich O. Beese
Agronomist: Director of the Institute of Soil Science and Forest Nutrition at the University of Göttingen (Institut für Bodenkunde und Waldernährung an der Universität Göttingen)

Prof. Gotthilf Hempel
Fishery biologist: Director of the Centre for Marine Tropical Ecology at the University of Bremen(Zentrum für Marine Tropenökologie an der Universität Bremen)

Prof. Paul Klemmer
Economist: President of the Rhenish-Westphalian Institute for Economic Research, Essen (Rheinisch-Westfälisches Institut für Wirtschaftsforschung in Essen)

Prof. Lenelis Kruse-Graumann
Psychologist: Specialist in "Ecological Psychology" at the Open University, Hagen (Schwerpunkt "Ökologische Psychologie" an der Fernuniversität Hagen)

Prof. Karin Labitzke
Meteorologist: Institute for Meteorology at the Free University Berlin (Institut für Meteorologie der Freien Universität Berlin)

Prof. Heidrun Mühle
Agronomist: Head of Department of Agricultural Lands at the Environment Research Centre Leipzig-Halle (Projektbereich Agrarlandschaften am Umweltforschungszentrum Leipzig-Halle)

Prof. Hans-Joachim Schellnhuber (Vice Chairperson)
Physicist: Director of the Potsdam Institute for Climate Impact Research (Potsdam-Institut für Klimafolgenforschung)

Prof. Udo Ernst Simonis
Economist:Department of Technology–Work–Environment at the Science Centre Berlin (Forschungsschwerpunkt Technik–Arbeit–Umwelt am Wissenschaftszentrum Berlin)

Prof. Hans-Willi Thoenes
Technologist: Rhenish-Westphalian Technical Control Board, Essen (Rheinisch-Westfälischer TÜV in Essen)

Prof. Paul Velsinger
Economist: Head of the Department of Regional Economics at the University of Dortmund (Fachgebiet Raumwirtschaftspolitik an der Universität Dortmund)

Prof. Horst Zimmermann (Chairperson)
Economist: Department of Public Finance at the University of Marburg (Abteilung für Finanzwissenschaft an der Universität Marburg)

German Advisory Council on Global Change

World in Transition:

The Research Challenge

Annual Report 1996

with 12 Illustrations

 Springer

GERMAN ADVISORY COUNCIL ON GLOBAL CHANGE (WBGU)
Secretariat at the Alfred-Wegener-Institute for Polar and Marine Research
Columbusstraße
D-27568 Bremerhaven
Germany

Acknowledgments:

External contributions and corrections to this report are gratefully acknowledged from

Dr. A. Becker, Potsdam Institut für Klimafolgenforschung
Dipl.-Pol. F. Biermann, LL.M., Wissenschaftszentrum Berlin
Dr. A. Bronstert, Potsdam Institut für Klimafolgenforschung
Prof. Dr. D. Cansier, Wirtschaftswissenschaftliche Fakultät der Universität Tübingen
Prof. Dr. A. Endres, Fernuniversität Hagen, Fachbereich Wirtschaftswissenschaften
Dipl.-Soz. A. Engels, Universität Bielefeld, Institut für Wissenschafts- und Technikforschung
Prof. Dr. H.-D. Evers, Universität Bielefeld, Institut für Soziologie
Prof. Dr. C. F. Gethmann, Universität GH Essen, Fachbereich Philosophie
Prof. Dr. H. Graßl, WCRP Sekretariat, Genf
Prof. Dr. J. H. Hohnholz, Institut für wissenschaftliche Zusammenarbeit mit Entwicklungsländern, Tübingen
Prof. Dr. V. Kreibich, Universität Dortmund, Fachgebiet Geographische Grundlagen
Dr. H.-J. Luhmann, Wuppertal-Institut, Abteilung Klimapolitik
Prof. Dr. J. Luther, Fraunhofer-Institut für Solare Energiesysteme, Freiburg
Prof. Dr. R. Müller, Max-Planck-Gesellschaft, Arbeitsgruppe Internationales Umweltrecht, Halle
Dr. S. Oberthür, Gesellschaft für Politikanalyse, Berlin
Dr. T. Plän, Institut für Naturschutzforschung, Regensburg
Prof. Dr. M. Pye, Universität Marburg, Fachgebiet Allgemeine und Vergleichende Religionswissenschaften
Prof. Dr. O. Rentz, Institut für Industriebetriebslehre und Industrielle Produktion, Karlsruhe
PD Dr. H. Schrader, Universität Bielefeld, Institut für Soziologie
Prof. Dr. W. Ströbele, Universität Münster
Dr. K. Urban, Potsdam Institut für Klimafolgenforschung
Prof. Dr. V. von Prittwitz, Gesellschaft für Politikanalyse, Berlin
Prof. Dr. J. Weimann, Universität Magdeburg, Fakultät für Wirtschaftswissenschaften
Prof. Dr. P. Weingart, Universität Bielefeld, Institut für Wissenschafts- und Technikforschung
Dr. P. Wiedemann et al., Forschungszentrum Jülich, Programmgruppe MUT
Dr. M. Winterhagen, Universität Bielefeld, Institut für Wissenschafts- und Technikforschung

The Council thanks the collaborators of the Core Project "Questions" (Potsdam Institut für Klimafolgenforschung) and of the BMBF-Project "Syndrome Dynamics" (No. 01LG9401-03)

Cataloging in Publication applied for
World in transition: the research challenge / German Advisory Council on Global Change. [Transl.: Tim Spence]. – Berlin ; Heidelberg ;
New York ; Barcelona ; Budapest ; Hong Kong ; London ; Milan ; Paris ; Santa Clara ; Singapore ; Tokyo : Springer, 1997
(Annual report / German Advisory Council on Global Change ; 1996) Dt. Ausg. u.d.T.: Welt im Wandel: Wege zur Lösung globaler Umweltprobleme

1996. World in transition: the research challenge. – 1997

ISBN-13: 978-3-642-64468-9   e-ISBN-13: 978-3-642-60587-1
DOI: 10.1007/978-3-642-60587-1

Translation: Tim Spence, Bremen
Cover design: E. Kirchner, Heidelberg using the following illustrations
Child with firewood, B. Pilardeaux, WBGU
Medical research project in Thailand: S. Esche, Gesellschaft für Technische Zusammenarbeit (GTZ) GmbH
Ozone profiler: J. Notholt, AWI
Research vessel Polarstern: AWI
Globe: IFA Bilderteam
SPIN 10553534 32/3137-5 4 3 2 1 0 - Printed on Workprint 100% recycling paper

Outline of Contents

Contents

Boxes

Tables and Figures

Acronyms

ACSYS	Arctic Climate System Study (WCRP)
AIDA	Arbeitsgemeinschaft der international ausgerichteten deutschen Agrarforschung [German Alliance for International Agricultural Research]
ALG	Alaska Latitudinal Gradient (IGBP)
APE	Airborne Polar Experiment (ESF)
APN	Asia-Pacific Network for Global Change
ATSAF	Arbeitsgemeinschaft Tropische und Subtropische Agrarforschung, Bonn [Working Group on Tropical and Subtropical Agricultural Research]
AWI	Alfred-Wegener-Institut für Polar und Meeresforschung, Bremerhaven [Alfred Wegener Institute for Polar and Marine Research]
BAHC	Biospheric Aspects of the Hydrological Cycle (IGBP)
BALTEX	Baltic Sea Experiment (GEWEX)
BAPMoN	Background Air Pollution Monitoring Network (WMO)
BfN	Bundesamt für Naturschutz, Bonn [Federal Office for Nature Conservation]
BFTCS	Boreal Forest Transect Case Study (IGBP)
BMBF	Bundesministerium für Bildung, Wissenschaft, Forschung und Technologie [Federal Ministry of Education, Science, Research and Technology]
BMI	Bundesministerium des Innern [Federal Ministry of the Interior]
BML	Bundesministerium für Ernährung, Landwirtschaft und Forsten [Federal Ministry of Food, Agriculture and Forestry]
BMU	Bundesministerium für Umwelt, Naturschutz und Reaktorsicherheit [Federal Ministry for Environment, Nature Conservation and Reactor Safety]
BMV	Bundesministerium für Verkehr [Federal Ministry of Transport]
BMWi	Bundesministerium für Wirtschaft [Federal Ministry of the Economy]
BMZ	Bundesministerium für wirtschaftliche Zusammenarbeit und Entwicklung [Federal Ministry for Economic Cooperation and Development]
BRIM	Biosphere Reserve Integrated Monitoring (MAB)
CDP	City Data Programme (UNCHS)
CENR	Commission on Environment and Natural Resources (USGCRP)
CGCP	Canadian Global Change Programme
CGER	Center for Global Environmental Research (Japan)
CGIAR	Consultative Group on International Agricultural Research
CIESIN	Consortium for International Earth Science Information Network (NASA)
CLIVAR	Climate Variability and Predictability Programme (WCRP)
CPO	Core Project Office
CRISTA	Cryogenic Infrared Spectrometer and Telescope for the Atmosphere (NASA)
CSD	Commission on Sustainable Development (UN)
CSEC	Centre for Study of Global Environmental Change (UK)
CSERGE	Centre for Social and Economic Research on the Global Environment (UK)
DecCen	Study of Decadal-to-Centennial Climate Variability and Predictability (CLIVAR)
DIVERSI-TAS	Ecosystem Function of Biodiversity Programme (SCOPE, UNESCO, IUBS)
DFG	Deutsche Forschungsgemeinschaft [German Research Foundation]

DIE	Deutsches Institut für Entwicklungspolitik GmbH, Berlin [German Institute for Development Policy]
DKRZ	Deutsches Klimarechenzentrum, Hamburg [German Climate Computing Center]
DSE	Deutsche Stiftung für Internationale Entwicklung, Berlin [German Foundation for International Development]
DVW	Development Watch (UNDP)
DWD	Deutscher Wetterdienst, Offenbach [German Weather Service]
EA	Environment Agency of Japan
ECOPS	Committee on Ocean and Polar Sciences (EU, ESF)
EERO	European Environmental Research Organisation
EFEDA	Echival Field Experiment in a Desertification Threatened Area (EU)
EISMINT	European Ice-sheet Modelling Initiative (ESF)
ENRICH	European Network for Research in Global Change (EU)
ENSO	El Niño-Southern Oscillation
ENVISAT	Environmental Satellite (ESA)
EOS	Earth Observing System (NASA)
EPA	Environmental Protection Agency (USA)
EPD	Environment and Population Education & Information for Development (UNESCO)
EPICA	European Polar Ice Coring in Antarctica (ESF)
ERS-2	European Research Satellite 2
ESA	European Space Agency
ESF	European Science Foundation
ESRC	Economic and Social Research Council (UK)
EU	European Union
EUREKA	European Research and Co-ordination Agency
EUROEN-VIRON	Environmental Protection Project (EUREKA)
EURO-TRAC	European Experiment on Transport and Transformation of Environmentally Relevant Trace Constituents in the Troposphere over Europe (EUREKA)
FAIR	Agriculture and Fisheries Programme (EU)
FAL	Bundesforschungsanstalt für Landwirtschaft, Braunschweig [Federal Research Institute for Agriculture]
FAM	Forschungsverbund Agrarökosysteme München [Munich Agricultural Ecosystem Research Association]
FAO	Food and Agriculture Organisation of the United Nations
FhG	Fraunhofer-Gesellschaft zur Förderung der angewandten Forschung, München [Fraunhofer Society]
FZK	Forschungszentrum Karlsruhe [Karlsruhe Research Center]
GAIM	Global Analysis, Interpretation and Modelling (IGBP)
GATT	General Agreement on Tariffs and Trade
GAW	Global Atmosphere Watch (GCOS)
GBA	Global Biodiversity Assessment (UNEP)
GBF	Gesellschaft für Biologische Forschung, Braunschweig [Biological Research Association]
GCDIS	Global Change Data and Information System
GCM	General Circulation Models
GCOS	Global Climate Observing System (WMO, IOC, UNESCO, UNEP, ICSU)
GCRIO	Global Change Research and Information Office (USA)
GCTE	Global Change and Terrestrial Ecosystems (IGBP)
GECP	Global Environmental Change Programme of ESCR
GEENET	Global Environmental Epidemiology Network (WHO, UNEP)
GEF	Global Environment Facility (UN)
GELNET	Global Environmental Library Network (WHO, UNEP)
GEMS	Global Environmental Monitoring System (UNEP)

GEMS-HEALS	Human Exposure Assessment Location Project (UNEP)
GENIE	Global Environment Network for Information Exchange in the UK
GERP	Global Environment Research Program of Japan
GETNET	Global Environmental Technology Network (WHO, UNEP)
GEWEX	Global Energy and Water Cycle Experiment (WCRP)
GFE	Großforschungseinrichtung (Helmholtz-Zentrum) [Helmholtz Research Centers]
GIEWS	Global Information and Early Warning System (FAO)
GIS	Geographical Information System
GISDATA	Geographical Information Systems: Data Integration and Data Base Design (ESF)
GKSS	Forschungszentrum Geesthacht [Geesthacht Research Center]
GLOBEC	Global Ocean Ecosystem Dynamics (USGCRP)
GLOSS	Global Sea Level Observing System (IOC)
GO$_3$OS	Global Ozone Observing System (WMO)
GOALS	Global Ocean-Atmosphere-Land System (CLIVAR)
GOAP	Greifswalder Bodden und Oderästuar-Austauschprozesse (LOICZ) [Land-Ocean Exchange Processes in the Greifswalder Bodden and Oder Estuary]
GOES	Global Omnibus Environmental Survey (IHDP)
GOME	Global Ozone Monitoring Experiment (EU)
GOOS	Global Ocean Observing System (WMO)
GRID	Global Resources Information Database (UNEP)
GRIP	Greenland Ice-core Project (ESF)
GSF	Forschungszentrum für Umwelt und Gesundheit, Neuherberg bei München [Research Center for Environment and Health]
GTOS	Global Terrestrial Observing System (WMO)
GTZ	Deutsche Gesellschaft für Technische Zusammenarbeit, Eschborn [German Association for Technical Cooperation]
HCP	Healthy City Project (WHO)
HD	Hokkaido Development Agency (Japan)
HEALTH	Maastricht Health Research Institute for Prevention and Care
HGF	Hermann-von-Helmholtz-Gemeinschaft Deutscher Forschungszentren, Bonn [Hermann von Helmholtz Association of German Research Centers]
HSD	Human Settlements Database (UNCHS)
HWRP	Hydrology and Water Resources Programme (WMO)
IACGEC	Inter-Agency Committee on Global Environmental Change (UK)
IAI	Inter-American Institute for Global Change Research (NSF)
ICLIPS	Integrated Assessment of Climate Protection Strategies (PIK)
ICSU	International Council of Scientific Unions
IDNDR	International Decade for Natural Disaster Reduction (UN)
IEA	International Energy Agency
IGAC	International Global Atmospheric Chemistry Project (IGBP)
IGBP	International Geosphere Biosphere Programme (ICSU)
IGBP-DIS	IGBP Data and Information System
IGBP-SAC	IGBP Scientific Advisory Council
IGFA	International Group of Funding Agencies
IGOSS	Integrated Global Ocean Service System (IOC, WMO)
IHDP	International Human Dimension of Global Environmental Change Programme (ICSU)
IHDP-DIS	IHDP Data and Information System
IHP	International Hydrological Programme (UNESCO)
IIASA	International Institute for Applied Systems Analysis (Österreich)
ILO	International Labour Organisation
IMA	Interministerielle Arbeitsgruppe der Bundesregierung [Interministerial Working Group]

Infoterra	International Environmental Information System (UNEP)
IOC	Intergovernmental Oceanic Commission (UNESCO)
IPCC	Intergovernmental Panel on Climate Change (WMO, UNEP)
IPK	Institut für Pflanzengenetik und Kulturpflanzenforschung, Gatersleben [Institute for Plant Genetics and Crop Plant Research]
IRPTC	International Register of Potentially Toxic Chemicals (UNEP)
ISI	Fraunhofer-Institut für Systemtechnik und Innovationsforschung, Karlsruhe [Fraunhofer Institute for Systems and Innovation Research]
ISRIC	International Soil Reference and Information Centre (Niederlande)
ISSC	International Social Science Council
IUBS	International Union of Biological Sciences (UNESCO, SCOPE)
JGOFS	Joint Global Ocean Flux Study (IGBP)
JPS	Joint Planning Staff of WCRP
JSC	Joint Scientific Committee of WCRP
KFA	Forschungszentrum Jülich [Jülich Research Center]
KNMJ	Royal Netherlands Meteorological Institute
KUSTOS	Küstennahe Stoff- und Energieflüsse (LOICZ) [Coastal Material and Energy Fluxes]
LBA	Large-scale Biosphere-Atmosphere Experiment in Amazonia
LME	Large Marine Ecosystems
LOICZ	Land-Ocean Interactions in the Coastal Zone (IGBP)
LTERM	Long-Term Environmental Research and Monitoring
LUCC	Land-Use/Land-Cover Change (IGBP, IHDP)
MAB	Man and the Biosphere Programme (UNESCO)
MADAM	Dynamik und Management von Mangroven (LOICZ)
MAFF	Ministry of Agriculture, Fisheries and Food (UK)
MAST	Marine Science and Technology Programme (EU)
MIPAS	Michelson Interferometric Passive Atmospheric Sounder (ESA)
MOST	Management of Social Transformation (UNESCO)
MPG	Max-Planck-Gesellschaft zur Förderung der Wissenschaften, München [Max Planck Society]
MPI	Max-Planck-Institut [Max Planck Institute]
MUT	Programmgruppe Mensch, Umwelt, Technik der KFK [Humanity, Environment and Technology Program Group at the Jülich Research Center]
NAFTA	North American Free Trade Agreement
NASA	National Aeronautics and Space Administration (USA)
NATO	North Atlantic Treaty Organisation
NATT	Northern Australian Tropical Transect (IGBP)
NDSC	Network for the Detection of Stratospheric Change Programme (EU)
NECT	North East Chinese Transect (IGBP)
NERC	National Environment Research Council (UK)
NGO	Non-Governmental Organization
NOAA	National Oceanic and Atmospheric Administration (USA)
NRP	Dutch National Research Programme on Global Air Pollution and Climate Change
NSF	National Science Foundation (USA)
NSTC	National Science and Technology Council (USA)
ODA	Overseas Development Administration (UK)
OECD	Organisation for Economic Co-operation and Development
OFP 2	Zweites Ozonforschungsprogramm der Bundesregierung [Second Ozone Research Program]
OHP	Operational Hydrology Programme (WMO)
ÖKOBOD	Ökosystem Boddengewässer - Organismen und Stoffhaushalt (LOICZ) [Organisms and Matter Cycles in Aquatic Ecosystems - Marine Inlets]
PAGEC	Perceptions and Assessment of Global Environmental Conditions and Change (IHDP)

PAGES	Past Global Changes (IGBP)
PEET	Partnerships for Enhancing Expertise in Taxonomy (USA)
PHARE	Pologne-Hongrie Assistance Pour la Restructuration [Polish-Hungarian Reconstruction Support]
PIK	Potsdam Institut für Klimafolgenforschung [Potsdam Institute for Climate Impact Research]
RIM	Regional Integrated Models
RIVM	Netherlands National Institute for Public Health and Environment
SALT	Savannas in the Long Term (IGBP)
SAP	Structural Adjustment Programme
SAUF	Senatsausschuß für Umweltforschung [Senate Committee for Environmental Research]
SC-IGBP	Scientific Committee of the IGBP
SCIA-MACHY	Scanning Imaging Absorption Spectrometer for Atmospheric Cartography (ESA)
SCOPE	Scientific Committee on Problems of the Environment (ICSU)
SDNP	Sustainable Development Networking Programme (UNEP)
SFB	Sonderforschungsbereich (DFG) [Collaborative Research Center]
SFS	Sciences for Food Security (ICSU)
SGCR	Subcommittee on Global Change Research (USA)
SHIFT	Studies on Human Impact on Forests and Foodplains in the Tropics (BMFT)
SI	Smithsonian Institution (USA)
SoE	State of the Environment (UNEP)
SPARC	Stratospheric Processes and their Role in Climate (WCRP)
SPPU	Schweizer Schwerpunktprogramm Umwelttechnologie und Umweltforschung [Swiss Priority Program Environmental Technology and Environmental Research]
SRU	Rat von Sachverständigen für Umweltfragen [Council of Environmental Experts]
SSG	Scientific Steering Group
STA	Science and Technology Agency (Japan)
START	Global Change System for Analysis, Research, and Training (IGBP)
SVAT	Soil-Vegetation-Atmosphere-Transfer Model
TEMPUS	Trans-European Mobility Programme for University Studies (EU)
TERM	Tackling Environmental Resource Management (ESF)
TERN	Terrestrial Ecosystem Research Network (Deutschland)
TFS	Troposphärenforschungsprogramm der Bundesregierung [German Troposphere Research Program]
TOGA	Tropical Ocean - Global Atmosphere (WCRP)
TRAINS	Trade Analysis and Information System (UNCTAD)
TRUMP	Transport- und Umsatzprozesse in der Pommerschen Bucht (LOICZ) [Matter Transport and Turnovers in the Pomeranian Bight]
TSER	Targeted Socio-Economic Research (EU)
TVA	Tennessee Valley Authority (USA)
UBA	Umweltbundesamt, Berlin [Federal Environment Agency]
UFOKAT	Umweltforschungskatalog (UBA) [Directory of Environmental Research]
UFZ	Umweltforschungszentrum Leipzig-Halle [Leipzig-Halle Research Center]
UMP	Urban Management Programme (World Bank, UNDP, UNCHS)
UN	United Nations
UNCED	United Nations Conference on Environment and Development
UNCHS	United Nations Centre for Human Settlements (HABITAT)
UNCTAD	United Nations Conference on Trade and Development
UNDP	United Nations Development Programme
UNEP	United Nations Environment Programme
UNEP-EAP	UNEP Environment Assessment Programme

UNESCO	United Nations Educational, Scientific and Cultural Organization
UNICEF	United Nations Children Fund
USAID	U.S. Agency for International Development
USGCRP	U.S. Global Change Research Program
WAVES	Water Availability, Vulnerability of Ecosystems and Society (joint German-Brazilian research project)
WBGU	Wissenschaftlicher Beirat der Bundesregierung Globale Umweltveränderungen [German Advisory Council on Global Change]
WBL	Wissenschaftsgemeinschaft Blaue Liste, Dortmund [Association of "Blue List" Research Centers]
WCAP	World Climate Assessment Programme (WMO)
WCDP	World Climate Data Programme (WMO)
WCIRP	World Climate Impacts and Response Strategies Programme (WCP)
WCP	World Climate Programme (ICSU)
WCRP	World Climate Research Programme (WMO, ICSU)
WDC	World Data Centres (ICSU)
WFP	World Food Programme (FAO)
WGMS	World Glacier Monitoring Service (UNEP)
WHO	World Health Organization (UN)
WMO	World Meteorological Organisation (UN)
WOCE	World Ocean Circulation Experiment (WCRP)
WRI	World Resources Institute
WTO	World Trade Organisation
WWW	World Weather Watch (WMO)
ZALF	Zentrum für Agrarlandschafts- und Landnutzungsforschung, Müncheberg [Center for Agricultural Landscape and Land-Use Research]

Summary

Introduction

For the first time in history, human activities are having impacts of planetary scale. The resultant changes in the global environment are reshaping the relationship between humankind and the natural basis on which its existence depends. This transformation process, called global change, is occurring at unprecedented speed and involves many risks. It can only be understood if Earth is conceived of holistically as a single system. Global change poses a major challenge for the scientific community, which must describe and explain how the Earth System is altered by human intervention, how these processes are influenced in turn by natural changes in the Earth System, and, finally, whether and to what extent there are ways to control global change.

Anthropogenic climate change is a good example for the sheer dimensions of human-induced effects on the global environment. Carbon dioxide emissions from transport in Germany contribute towards rising sea level and the expected disappearance of coral islands 20,000 km away, thus robbing the inhabitants of such islands of their very habitat. Humankind is thus confronted not only with a major ethical dilemma, but also with a complex of difficult research issues and problems which have to be resolved as quickly and as competently as possible. These problems can only be solved by interdisciplinary and international research networks in which, alongside climate modeling and hydrography, for example, other disciplines such as the philosophy of law and cultural anthropology must also play a role.

Global change research poses an enormous challenge for the researchers themselves, for the bodies funding and promoting such research as well as for those who base their decision-making on its findings, and demands integrative capacity, flexibility and imaginative power from scientists, research promoting organizations and users. Innovative guidelines and structures are necessary for handling the complex problems and for developing the problem-solving competence. "Traditional" environmental research has so far proved inappropriate to meet these demands.

In its 1993, 1994 and 1995 Annual Reports, the Council identified and described the core problems associated with global change – on the one hand, the changes in people's natural environment (the "ecosphere"), on the other, the changes with society itself (the "anthroposphere"). The 1996 Annual Report now focuses on the organization of global change research and examines the conditions for enhancing Germany's contribution.

Suspicions have often been voiced that calls for more research serve only to divert attention from the necessity of environmental action. However, research explicitly geared to solving specific problems is indeed relevant for political action, improving the decision-making competence on which any such action depends.

The Council's focus in this year's Annual Report embraces not only the "classical" fields of environmental research within the natural sciences, but also the economic and sociocultural dimensions of global change. The methodological foundations were established in previous Reports through the development of an integrative research approach based on the analysis of syndromes (WBGU, 1993 and 1994). This approach permits the operationalization of the networked thinking that is essential for mitigating global change and offers new options for the design and organization of global change research.

When public funds are scarce, clear-cut priorities are required in order to achieve maximum efficiency in the selection and execution of research projects. To this end, the Council has elaborated relevance criteria and integrational principles for global change research. These criteria and principles could also be applied in the current reshaping of Germany's environmental research programs. The Council welcomes the interministerial initiative of the *Bundesministerium für Bildung, Wissenschaft, Forschung und Technologie* (BMBF) (Federal Ministry for Education, Science, Research and Technology) and the *Bundesministerium für Umwelt, Naturschutz und Reaktorsicherheit* (BMU) (Federal Ministry for Environment, Nature Conservation and Reactor Safety) for a new environmental research program to succeed several specialized programs and government support frameworks for global change research in the fields of climate, marine, polar and ozone research.

In this year's Report, the Council develops guidelines for the essential restructuring and reorientation of selected areas within Germany's global change research, while taking established structures into consideration. This entails, on the one hand, a description of integrated research approaches, as illustrated by a case study, and, on the other, a survey of the classical sectors within global change research, an evaluation of Germany's involvement in international research programs, and the identification of research gaps in specific sectors.

Integrated Global Change Research

Systems Approach

A cardinal feature of global change is that humankind itself is now an active factor within the Earth System, playing a significant role at the planetary scale. Human interventions in that system, as manifested in the depletion of raw materials, shifts in material and energy fluxes, changes to large-scale natural structures and critical stresses on environmental assets, are altering the very character of the Earth System to an increasing degree. The complexity of these processes poses a major challenge for the scientific community and generates a number of new research issues, listed below. Finding answers to these questions will be of increasing importance in the years to come:

- How are changes in the ecosphere produced and how are these linked to global development problems?
- How can they be identified or even predicted at an early stage?
- What risks do they involve?
- How must humankind act in order to prevent negative developments at the global level, to avert threats and/or mitigate the consequences of global change?

Such research should be guided by the principle of sustainable development. The crucial element of this concept, now generally acknowledged, is the interdependence of environment and development (AGENDA 21). This reflects a growing insight that human beings and their environment are closely integrated within a system of mutual interaction. Research on global change is therefore confronted with two fundamental problems. Firstly, the investigation of the Earth System requires an integrative approach because the interactions between its components operate across the boundaries of single disciplines, sectors or environmental media. The second fundamental problem is the enormous complexity of the dynamic interrelationships involved, which makes a distinct description, any overall analysis and modeling much more difficult. The only approach capable of responding adequately to these problems is one that is networked and interdisciplinary. The sectoral bias within research must be supplemented by a systems approach that establishes cross-linkages between different strands of research.

The Council has proposed a new method for holistic analysis of the present crisis of the Earth System (WBGU, 1993 and 1995a). The elements chosen for that analysis are not, as is often the case elsewhere, a set of easily indexed base variables, such as atmospheric concentrations of CO_2, size of population or gross national product. Instead, the most important global trends are being used as qualitative elements. They are termed trends of global change and provide information about the dominant features of global development. The development of the Earth System is then described using this set of trends. Although highly complex natural or anthropogenic processes are being analyzed, the internal processes that make up their detail are not lost sight of.

Those trends possessing special relevance for global change are selected on the basis of educated guesses. They are not evaluated at the outset; problematic processes such as climate change, loss of biological diversity or soil erosion are placed alongside other trends – like globalization of markets, or progress in biotechnology and genetic engineering – which can have positive or negative impacts depending on the perspective taken and the specific manifestation of the trend in question. Another category of trends are those which may lead to the mitigation of global problems, e.g. strengthening of environmental protection efforts at national level, growing environmental awareness, or the growth of international regimes.

The various trends and their interactions can be combined in a qualitative global network of interrelations, which describes global change as a system and which represents the starting point for more extensive analysis of the Earth System's dynamics. With the help of this empirical-phenomenological description of global change, it is possible to design qualitative models, the subject of a current BMBF research project.

Syndrome Concept

Networks of interrelations can be developed for other levels besides the global. A regionalized analysis of the Earth System using this instrument provides clear indication that the interactions in certain regions between human societies and the environment frequently operate according to typical patterns. These functional patterns (syndromes) are unfavourable and characteristic constellations of natural and civilizational trends and their respective interactions, and can be identified in many regions of the world. The Council's underlying thesis is that complex global environmental and development problems can be attributed to a discrete number of environmental degradation patterns.

Syndromes are transsectoral in nature; while specific problems may affect several sectors (such as the economy, the biosphere, population), they are always related, directly or indirectly, to natural resources. Syndromes are globally relevant when they modify the Earth System and have a noticeable impact, directly or indirectly, on the basis of life for a major part of humankind, or when global solutions are needed to surmount the problems. This year's Report includes an attempt at identifying the globally significant syndromes under which Planet Earth is suffering (*see box "Overview of Global Change Syndromes"*).

The syndrome concept provides a new basis for global change research, which continues to be organized according to the environmental media or core problems. Given the desiderata for global change research – interdisciplinarity, internationality and problem-solving competence – it is obvious that future environmental research should be structured along transdisciplinary lines. In this connection, the Council's syndrome concept offers new options for shaping research activities. It is therefore recommended that these syndromes be adopted as the central objects of future global change research.

Relevance Criteria and Integration Principles

The importance of global change for the future development of humankind, and the uniqueness, complexity, variety and dynamics of the phenomena involved, make it necessary to deploy a number of additional relevance criteria for research policy. Putting these criteria into operation can fulfill a dual purpose – orienting research activity to the cross-sectional character of environmental issues, and achieving more efficient prioritization when financial resources are scarce. The Council's recommendation is that the following criteria in particular be applied when selecting research topics in the field of global change:

- *Global relevance:* Are key parameters, basic patterns or core problems in the Earth System being investigated? Are large numbers of people affected by the problem? Is the research likely to generate new options for controlling the environment/development process?
- *Urgency:* Are answers needed quickly in order to prevent irreversible environmental or socioeco-

---

BOX

Overview of Global Change Syndromes

"UTILIZATION" SYNDROMES

1. Overcultivation of marginal land: *Sahel Syndrome*
2. Overexploitation of natural ecosystems: *Overexploitation Syndrome*
3. Environmental degradation through abandonment of traditional agricultural practices: *Rural Exodus Syndrome*
4. Non-sustainable agro-industrial use of soils and bodies of water: *Dust Bowl Syndrome*
5. Environmental degradation through depletion of non-renewable resources: *Katanga Syndrome*
6. Development and destruction of nature for recreational ends: *Mass Tourism Syndrome*
7. Environmental destruction through war and military action: *Scorched Earth Syndrome*

"DEVELOPMENT" SYNDROMES

8. Environmental damage of natural landscapes as a result of large-scale projects: *Aral Sea Syndrome*

9. Environmental degradation through the introduction of inappropriate farming methods: *Green Revolution Syndrome*
10. Disregard for environmental standards in the course of rapid economic growth: *Asian Tigers Syndrome*
11. Environmental degradation through uncontrolled urban growth: *Favela Syndrome*
12. Destruction of landscapes through planned expansion of urban infrastructures: *Urban Sprawl Syndrome*
13. Singular anthropogenic environmental disasters with long-term impacts: *Major Accident Syndrome*

"SINK" SYNDROMES

14. Environmental degradation through large-scale diffusion of long-lived substances: *Smokestack Syndrome*
15. Environmental degradation through controlled and uncontrolled disposal of waste: *Waste Dumping Syndrome*
16. Local contamination of environmental assets at industrial locations: *Contaminated Land Syndrome*

nomic developments with severe negative outcomes?

- *Gaps in knowledge:* Can serious gaps concerning a holistic view of the global environment and its dynamics be closed?
- *Responsibility:* Are problems being investigated for which Germany is directly or indirectly responsible (e.g. through greenhouse gas emissions or as participants on the world market)? Does the topic relate to general ethical principles (e.g. preservation of life on Earth)?
- *National impact:* Are problems being researched which could have direct or indirect effects on Germany (e.g. impacts on climate, environmental refugees)?
- *Research and problem-solving competence:* Does the research relate to areas where Germany makes a substantial contribution on account of its scientific, technological and infrastructural potential? Can research on the topic lead to further improvement of that potential and thus to enhancement of Germany's attractiveness for investment?

Since it is neither reasonable nor feasible for German global change research to cope with all syndromes simultaneously, priorities should be set with the help of the above criteria. Moreover, investigation of the various syndromes should be pursued by the international scientific community. A survey conducted within the WBGU on the basis of the relevance criteria produced an initial ranking of the syndromes. Seven syndromes were given uppermost priority (listed in alphabetic order):

- *Contaminated Land Syndrome*
- *Dust Bowl Syndrome*
- *Mass Tourism Syndrome*
- *Sahel Syndrome*
- *Smokestack Syndrome*
- *Urban Sprawl Syndrome*
- *Waste Dumping Syndrome*

The specific recommendations are:

- to discuss and improve the syndrome concept in a series of symposia involving research scientists and decision-makers from different sectors of society. The current list of syndromes may undergo further modifications in the process;
- to produce an improved ranking of the syndromes based on a Delphi study;
- immediately establish three research networks among existing establishments for pilot studies of the *Smokestack, Sahel* and *Urban Sprawl syndromes.* These integrated studies could function as key projects under the Federal Government's new Environmental Research Program.

Achieving a global perspective requires collaboration between and the integration of different disciplines, interest groups and actors. The diversity of concepts for communicating environmental knowledge means that many problems must be overcome for such integration to occur. The key issue for researchers concerns the principles according to which the requisite synthesis is to be realized. Simply calling for "networking", "interdisciplinarity" or "interaction" is inadequate as an approach – what is needed are principles and instruments providing a concrete basis for the holistic analysis of global change syndromes.

In this year's Annual Report, the WBGU puts forward a number of principles that may prove helpful in the implementation of integrated environmental research (*integration principles*). These principles relate to analytic, methodological and organizational aspects, as well as certain implementation aspects.

The Problem-Solving Process

Research on decision-making processes in the field of environmental policymaking has mainly been concerned with problems of national environmental policy. Although this has led to findings that have a bearing on the environmental decision-making process in the international and global framework as well, the situation is more complex. Global problems tend to be long-term in nature, which gives rise to major problems with regard to diagnosis and forecasting. This results in demands on early-warning systems and planning instruments, as well as on research methods and instruments. Global problems are also much more complex than environmental problems at the purely national level, with all the implications for the necessary consensus-formation. Conflicts are more difficult to resolve in an international context due to differences in culture, religion and level of development, the latter especially.

Research methods and approaches guided by national environmental policy must therefore be adapted in such a way that they can also be applied to the elements of the decision-making process relating to global environmental change. The focus should not be restricted to specific disciplines. Rather the various elements constituting the problem-solving process should be structured first. The next step is to ask which disciplines have already contributed or which disciplines should make a greater contribution to interdisciplinary research in future.

The next step is to ascertain which results are already available and in what respects they require supplementing. Distinction can be drawn between the following elements of problem-solving processes:

- Initial treatment of the problem. The starting point for solving global change problems is analysis, i.e. the identification of causes and effects and the assessment of future trends (forecasts). Due to the complexity of the subject and the need for integrated research approaches, it is necessary to have an adequate methodology, such as systems analysis, for describing and explaining the problem and arriving at forecasts.

- Guiding principles and objectives. Once the problem has been analyzed, it is necessary to define guiding principles and objectives. In the view of the Council, there are serious gaps in research into guiding principles, which needs to be oriented towards the concept of sustainable development and made more specific by means of relevant principles for action and appropriate indicators.

- Responsible bodies. Policies for influencing global environmental change require appropriate bodies at various levels (global, regional, national and local). Because sovereign states are the bodies which take action at international level, special attention must be focused on the mechanisms of decision-making and action which take place there. The specific constellation of responsible bodies and the problem of how to achieve effective cooperation between these bodies must therefore be studied in greater detail. Suitable methods for such research have to be selected and/or refined, game theory being one example in this context.

- Instruments. Agreement on objectives is achieved using instruments which are either already available to global environmental policymaking or which have yet to be developed. Research into the strength and effectiveness of these instruments is needed and must be advanced further. In particular, research is needed on transsectoral instruments, such as the various international environmental conventions, but also on the various subordinate instruments which operate under those conventions.

- Implementation. Once international conventions have been adopted, the next question concerns their implementation and enforcement as well as options for imposing sanctions. In view of the fact that problem-solving processes often stagnate at precisely this stage, the obstacles which arise in this context must be subjected to precise analysis.

- Research into decision-making and risk management processes. In addition to research into the above elements of the decision-making process, especially the problem of responsible bodies and the effectiveness of policy instruments, it is essential to conduct research into decision-making and risk management, which involves investigating two specific features of the problem-solving process, namely the problem of consensus-formation in cases where there are fundamental disparities

of interest, and the question as to how to handle uncertainty.

Sectoral Research on Global Change

In this chapter, the Report describes the current status of German sectoral research on global change, including international involvement, and identifies the research gaps.

CLIMATE AND ATMOSPHERE RESEARCH

The advanced level of German research in this field must be maintained through continuous improvement of the existing infrastructure. German climate research, for example, occupies a leading position in the world in the development of coupled ocean-ice-atmosphere models, thanks to consistent support from the BMBF, the *Max-Planck-Gesellschaft* (MPG) (Max Planck Society) and the *Deutsche Forschungsgemeinschaft* (DFG) (German Research Foundation). This position can only be maintained through adequate human resources policy, continuous modernization of computing capacities and constant refinement of models. Research tasks of special relevance to global change are:

- Further development of coupled ocean-ice-atmosphere models for predicting climate along different spatial and temporal scales, and of integrated models for climate impact research.

- Research into the Earth's paleoclimate using ice cores and marine and limnetic sediments. There is a general lack of data from tropical regions and the Southern Hemisphere in this field.

- Continuation and/or commencement of measurements of the composition of the atmosphere (various characterizing substances) at selected stations in Germany and northern Europe (stratosphere monitoring), as well as at sea and in the tropics (troposphere monitoring) in the context of international monitoring programs.

- Systematic analysis of existing data from different parts of the atmosphere in order to improve our understanding of climate variability.

- Development and evaluation of satellite experiments for measuring parameters and trace gases of relevance for the climate.

- Investigating the influence of aerosols and clouds on climate.

- Experimental studies of tropospheric chemistry at low latitudes (using research aircraft).

Climate and atmosphere research, in the narrower sense, is conducted primarily by natural science. Research on the impacts of global change (especially climate impact research) should involve the human sciences as well. What is needed is:

- greater development of Regional Integrated Models (RIM),
- organization of transdisciplinary and transinstitutional research networks for studying issues of sectoral and political relevance.

HYDROSPHERE RESEARCH

As with climate research, it is essential that the advanced level of German research in this field be maintained through continuous improvement of the existing infrastructure. A committed role in the Joint Global Ocean Flux Study (JGOFS), Global Ocean Ecosystem Dynamics Programme (GLOBEC) and Land-Ocean Interactions in the Coastal Zone Programme (LOICZ), all core projects of the International Geosphere Biosphere Programme (IGBP), is essential. Research tasks of special relevance to global change are:

- Development of the scientific foundations for an operational Global Ocean Observing System (GOOS).
- Research into human-induced impacts on marginal seas and coastal zones, and the development of the scientific basis for integrated management of coastal regions.
- Research into the polar oceans, with special reference to climatological aspects.

With regard to the global aspects of water resources, there is an enormous need for research into the causal interlinkages between climate, vegetation and anthroposphere, and for the development of environmentally sound land management practices that ensure water resources in the long term, as envisaged by the Land-Use and Land-Cover Change Programme (LUCC) and Biospheric Aspects of the Hydrological Cycle Programme (BAHC), core projects of IGBP.

Freshwater is a vital resource in all areas of life and society, functioning as a nutrient, a cultural asset and a production factor simultaneously. The WBGU considers it extremely important that research into freshwater resources be intensified. Research tasks of special relevance to global change are:

- Research into the conditions for increasing the supply of water for a growing world population.
- Research into the conditions for thrifty and sustainable use of water resources, in the sense of careful management of water resources in the various sectors of use (agriculture, industry, private households) and the equitable sharing of available water (intra- and intergenerational equity).
- Research into the conditions for preventing the pollution of surface water and groundwater stocks.

The main focus here is to develop dynamic models of the regional and global water balance, including feedback effects to the climate system, the biosphere and the anthroposphere.

SOIL RESEARCH

Soil research focuses primarily on the local and regional level, but it must now integrate global changes in climate, water balance and land use. The following fields are especially important in this connection:

- Quantification of soil functions: as storage media in the biogeochemical cycles of carbon, nitrogen, and sulfur, as well as the trace gas compounds of these elements which are responsible for climate forcing. Assessment of the potential influence on transformation processes exerted by changes in climate and land use.
- Degradation of soils due to the decoupling of element cycles through utilization. Impacts on the productivity and sustainable utilization of soils, and on the stability of recipient systems. Research activities at local, regional and national level.
- Effects of particulate substances removed from soils (through erosion) on the biotic components of neighboring limnetic and marine ecosystems (main focus on rivers, coral reefs and mangroves).
- Intensified use of remote sensing for Earth observation, and of computer simulation techniques for describing changes in terrestrial ecosystems at regional and global level.

BIODIVERSITY RESEARCH

Biodiversity, as a dimension of global change, is of such importance for the functions, stability and development of ecosystems that the Council considers it central to its recommendations. German biodiversity research still tends to focus too much on single disciplines and the purely national level. Wider conceptual frameworks and interdisciplinary links between the biosciences and the human sciences are still under development. The Council recommends that research be focused on the following areas:

- The basis for any assessment, preservation or restoration of biodiversity is a modern taxonomy that utilizes the methods and techniques of molecular biology more intensively, including advanced information technology. Research and educational facilities in this field are urgently in need of expansion if German researchers are to be involved in international biodiversity inventory projects and biogeographical assessments of biodiversity.
- Research should also focus on the dichotomy between the conservation and utilization of terrestrial and aquatic ecosystems. In particular, the interrelationships between diversity, stability and performance of ecosystems must be analyzed more intensively. Expanding research on population biology in order to improve nature conserva-

tion activities plays an important role in this connection. This calls for new approaches going beyond the all-too-narrow focus on biotope and species protection.

- High priority must be assigned to research into the impacts of environmental changes of varying quality, intensity and speed on populations, ecosystems and ecosystem functions (such as biogeochemical cycles). Such work should be based on findings from the areas specified above.
- Another important research field concerns the political efforts of the international community for the conservation and sustainable use of biodiversity. Research into the economics of biodiversity and the design of international environmental agreements is urgently needed.

POPULATION, MIGRATION AND URBANIZATION RESEARCH

Population trends, migration and urbanization are key factors in the analysis and management of global environmental problems. Population growth and poverty are powerful driving forces behind an overall trend that is now affecting industrialized countries as well, primarily in the form of mounting migrational pressure. Research in Germany is still inadequate, with respect to theoretical foundations, empirical case studies and simulation models, to analyze, forecast and respond to these trends. The following topics should be focused on:

- Rural-urban-relations must be re-investigated and re-assessed, taking into account the transfers between urban areas and the subsistence economy of surrounding rural areas (reversal of the push-pull approach).
- Identifying potential sources of migration and migration flows is an increasingly important task for international migration research. In particular, systematic research must be conducted into the motivational factors driving migration.
- The determinants influencing the individual's decision to migrate must be identified in terms of sociocultural nexus and the private household context. Traditional flow analysis must be enhanced through migration system research.
- Malnutrition, undernourishment and famine are major causes of migration. Research on food security and water availability must therefore be intensified.
- The informal economy plays a central role in providing a minimum level of social security for the urban poor. In-depth research into the development potential of this sector is therefore essential.
- Our knowledge about the increasing number of megacities and large-scale agglomerations and how these operate within the global system is still

fragmentary. There is also a lack of research on the informal growth of cities. To understand how "unplanned" megacities function, it is necessary to examine the systemic interrelation of these urban structures. The Second UN Conference on Human Settlements (HABITAT II) showed that the creation of adequate housing is an acute problem affecting the welfare of more than one billion people. Policy-oriented research should also be conducted in connection with international conferences (preparation and follow-up).

ECONOMICS RESEARCH

The Council sees a need for global economics research in the following three fields:

- Research on the objectives and impacts of global environmental policy. A key focus here should be the operationalization of the sustainable development principle. Above all, this requires the identification of the essential, i.e. non-renewable elements of natural capital, the assessment of the costs of neglected environmental protection, the evaluation of intra- and intergenerational distribution, especially the scientific debate on "correct" discounting methods, and the specification of criteria regarding the economic and social compatibility of sustainable development.
- Research on agencies responsible for global environmental policy. A principal research focus should be the economic analysis of the behavior of globally relevant actors – both political and private (such as multinational corporations). One major issue concerns the development of strategic behavioral options which produce benefits for the overwhelming majority of those involved.
- Research on the instruments of global environmental policy. Due to the limited planning, regulatory and fiscal options at the global level, environmental regimes are usually implemented through treaties, conventions and economic incentives. Research on economic instruments should therefore concentrate on further development of the tradable quotas/permits option (including joint implementation), the law of liability and global funds. Another issue, parallel to these, is the question of sanction mechanisms to be applied when treaties or conventions are violated by one or more party.

RESEARCH ON SOCIETAL ORGANIZATION

Research by political science on environmental topics has mainly focused on the national level, so it must now adopt a more global perspective. The problems experienced by newly industrializing nations and their growing importance for global change deserve special attention. Policy concepts relating to global environmental protection must also take into account the sociocultural and economic conditions and international law.

International environmental research must widen its focus to embrace not only global climate issues but also other problems such as soil degradation, loss of biodiversity, and the scarcity and contamination of water resources. In view of the discrepancy between environmental awareness and the policies which are actually implemented, analysis must center, as a matter of priority, on the process of political will-formation and the implementation of international treaties. Political research must also dedicate attention to the prevention of environmental conflicts. The following tasks need to be accomplished:

- Investigation of the socioeconomic, political and cultural restrictions on action and problems related to the implementation of international environmental treaties.
- Development of concepts on which to base problem-solving strategies for overcoming the typical obstacles encountered in global problem-solving processes (global commons, compliance issues, etc.).
- Analysis of the functional operation of international negotiation systems, with special reference to the uncertainty factor in global environmental change. Further, it is necessary to develop concepts for decision-making under uncertain premises.

Specialists in environmental law are examining the options for adopting and enforcing effective measures relating to global change. The legal issues include restricted national sovereignty, customary international law and ecological solidarity. Against this background, the Council recommends that the following legal issues be tackled:

- Clarification of the current body of extra-treaty standards and international customary law relating to global environmental problems, in order to react more flexibly to the latter.
- Defining a general obligation of ecological solidarity on the part of industrialized countries vis-à-vis the developing world.
- Clarification of the status of non-governmental organizations in international law.
- Clarification of legal issues concerning damages caused by global environmental change.
- Further development of enforcement mechanisms, decision-making procedures and dispute-settlement procedures in connection with international treaties.

RESEARCH ON THE PSYCHOSOCIAL SPHERE

The scientific disciplines covering the psychosocial sphere are devoting increasing attention to important issues in the analysis of the causes and effects of global change and interventions to remedy the problems which exist. This research is still little developed in Germany, with most projects involving only one discipline and decentralized organization. The following topics should be focused on, preferably by joint projects:

- Research into guiding principles of sustainable development, from basic ethical principles to operationalization and empirical analyses.
- Studies on the determinants of behavioral traits relevant to global change (perception and assessment of global change phenomena, motivation, etc.) and on strategies for modifying behavior.
- Investigation and evaluation of interventions (specific contexts and groups of actors), in terms of the interactions between technical, economic, legal and psychosocial measures.
- Development, systematic application and evaluation of global change education.
- Development and establishment of a worldwide, comprehensive system of social monitoring (analogous to environmental monitoring).

Accomplishing these tasks requires more cultural and cross-cultural comparative research on social actors in the form of comprehensive, transdisciplinary case studies, whereby studies should be conducted across a wide range of spatial and temporal scales.

TECHNOLOGICAL RESEARCH

Technological research is a key factor in managing problems of global change. A prime example is the further development of energy technologies aimed at an environmentally, economically and socially acceptable energy mix. The main focus should be placed on researching and developing different energy options, including:

- Research on solar photovoltaics.
- Research on the utilization of wind energy, especially in developing countries.

In addition, the Council recommends the promotion of research programs on the impact of air transport on climate, and the development of air transport along environmentally acceptable lines. At the interface between technology and economics, the Council proposes the following research topics:

- Examination of the appropriateness and effectiveness of jointly implemented activities to reduce greenhouse gases.
- Development of cost-efficient reduction strategies for greenhouse gas emissions, taking the radiatively active trace gases into account.
- Research concerning techniques for removing and storing CO_2, with special reference to ecological and economic aspects.
- Analysis and quantification of the impacts of greenhouse gas emissions on the emissions of oth-

er mass contaminants of the atmosphere and other environmental problems.

- Development of cost-efficient strategies for reducing tropospheric ozone.
- Development of logistics-oriented production processes (e.g. reduction of transport within the production process).
- Identification of environmentally sound industrialization processes in developing and newly-industrializing countries, taking into consideration the local technology and human resource potentials.

Practical, technology-based solutions to complex environmental problems require cooperation between various disciplines, depending on the respective project and the specific problem it addresses. The following fields have a role to play in this context:

- Technologies: engineering science.
- Materials: chemistry, biology and geology.
- Planning and design: economics and social sciences.
- Applications and effects: social and behavioral sciences, environmental medicine.

The Organization of Research

German research must undergo major structural improvements if it is to meet the needs of modern global change research. These include improvements to existing institutions, incentives for innovative research projects, especially in tertiary education, and enhanced coordination of research and research promotion. Demands for greater investment in research are frustrated by the scarcity of public funds. Lack of finance is a major obstacle, blocking further growth in research personnel and equipment budgets and, through non-selective staff cuts, deprives research institutions of opportunities to explore new research pathways. Shortage of public funds has imposed a restrictive framework that must be taken into consideration whenever organizational recommendations are made. The research community is therefore compelled to think about structural changes which might generate improved efficiency. Nevertheless, for all the problems that exist, Germany's research organization has many advantages.

The strengths of a federal and pluralist structure, and the number and variety of research institutions this entails, stems from the fact that individual groups can tackle new issues flexibly and choose their own partners, especially when scientific encouragement or financial incentives are provided. On the other hand, this structure is highly intricate, which in turn hinders the concentration of research capacities

under one central topic and the execution of long-term projects within international programs.

The *Wissenschaftsrat* (German Science Council) has given attention to these problems and has drawn up a set of recommendations for transdisciplinary environmental research at German universities, polytechnics and other research establishments. The obstacles are even greater for global change research, however, on account of the international context and the need to carry out investigations with foreign partners. This also explains why, in certain areas of global change research, German involvement in international programs and cooperation with developing countries is relatively confined.

Against this background, the Council puts forward a number of general organizational recommendations, grouped under three headings:

- Strengthen existing facilities and utilize approved instruments.
- Create new facilities.
- Coordinate the promotion of research.

Strengthen Existing Facilities and Utilize Approved Instruments

Existing research establishments must be given the capacity to continue ongoing projects in the field of global change research and/or to relate projects to global problems, and to start new projects involving cooperation at national and/or international level. This recommendation is directed at universities and polytechnics and to extra-university research establishments such as the *Max-Planck-Gesellschaft* (MPG) (Max Planck Society), the *Helmholtz-Gemeinschaft Deutscher Forschungszentren* (HGF) (Helmholtz Association of German Research Centers), the *Wissenschaftsgemeinschaft Blaue Liste* (WBL) (Association of "Blue List" Research Centers) and the *Fraunhofer Gesellschaft* (FhG) (Fraunhofer Society), as well as the research facilities operated by certain federal agencies. Impulses in this direction must come from the facilities themselves or from the bodies which operate and control them, i.e. by redefining the priorities and content of research and by organizational restructuring.

What is absolutely essential, however, is the use of approved support instruments on the part of the BMBF (joint projects, research networks) and the DFG (priority programs, collaborative research centers). Research groups and graduate colleges are suitable instruments, whereby the restrictive principle that research units must be located in a single institution should be relaxed in view of the technical opportunities provided by modern telecommunications.

All these integrating measures should also be applied in the education and training of domestic and foreign students and prospective scientists. Aspects of global change should be referred to during basic level courses, and studied in greater detail in advanced courses.

Major items of research equipment are absolutely essential in many areas of global change research. These include equipment for remote sensing and climate research using supercomputers, research vessels, remote sensing satellites and monitoring stations. Global change research also needs large-scale, comprehensive and long-term observation data on the environment, the economy and sociocultural aspects. It relies on comparisons between cultures and ecosystems and must build on detailed and broadly conceived case studies as well as complex models.

The Council attaches considerable importance to ensuring continued provision of these basic requirements. Germany's participation in international programs varies in quality, and in some important areas is in need of expansion. The Council recommends continued involvement in international institutions and secretariats, in terms of input, staffing and financial contribution, whereby greater integration of German researchers by such institutions would be desirable.

Create New Facilities

The Council recommends the establishment of a *German Strategy Center on Global Change* in order to enhance problem-solving capacity with respect to global change and to strengthen interdisciplinary cooperation. The Center would carry out complex problem analyses, using external experts as well to provide scientific support for decision-making processes. It should take up suggestions of policymakers and the public to translate these into research issues, as well as process existing scientific knowledge to support decision-making processes in politics, industry and society in general.

Small research centers should be set up at or around universities for limited periods of time; these would work on acute problems in the field of global change research over 10 years or so, and ensure German participation in international programs.

In addition, the Council recommends the creation of *research networks* as long-term, purpose-made alliances between independent research institutions for joint work on complex issues, such as a specific syndrome, and for further refinement of methodologies. These should include the use of modern technologies for data acquisition, storage and transmission within national and international frameworks. Re-

sponsible research bodies (MPG, HGF, WBL and FhG), the DFG, the BMBF and specialized research establishments and university departments should jointly create flexible institutions to deal with specific global change problems (inter-institutional research).

The Council appeals to industry and commerce, especially the multinational corporations, to set up a *Global Change Foundation* as an expression of environmental self-commitment. Such a body would help compensate for the financial restrictions referred to earlier. The Foundation should promote a dialog between the scientific community, economic policymakers and the media on global change issues. It could also be present at the EXPO 2000 World Exhibition in Hanover, Germany.

Coordinate the Promotion of Research

The two most important institutions providing funding and support for research in Germany are the BMBF and the DFG. The BMBF has several ministerial departments and various project support units responsible for specific research fields relating to global change. The same is true of the DFG, which is organized according to scientific disciplines. Both institutions must strengthen their efforts towards transdisciplinary planning and assessment. There is also a need for closer coordination between the DFG and the BMBF regarding the deployment of instruments for promoting global change research.

Within the Federal Government, supervisory control of global change research is not confined to the BMBF. Although the BMU does not operate its own research establishments, it supports a number of global change research projects through the *Umweltbundesamt* (UBA) (Federal Environmental Agency). Major research facilities and projects are also operated by the *Bundesministerium für Verkehr* (BMV) (Federal Ministry of Transport), *Bundesministerium für Wirtschaft* (BMWi) (Federal Ministry of Economy), *Bundesministerium für Ernährung, Landwirtschaft und Forsten* (BML) (Federal Ministry for Food, Agriculture and Forestry), *Bundesministerium für wirtschaftliche Zusammenarbeit und Entwicklung* (BMZ) (Federal Ministry of Economic Cooperation and Development) and *Bundesministerium des Innern* (BMI) (Federal Ministry of the Interior). The Council states a need for coordination here, which should go beyond the work of the *Interministerielle Arbeitsgruppe* (IMA) (Interministerial Working Group) on Global Change.

The Council is monitoring with great interest the efforts of the DFG to establish a *Nationales Komitee für Global Change Forschung* (German Global

Change Committee), comprising functional units of the *Senatsausschuß für Umweltforschung* (SAUF) (Senate Committee for Environmental Research) and the German IGBP Committee, for the purpose of planning and supporting research involvement in international global change programs. This National Committee could also play a role in coordinating the various global change research activities in Germany.

The Council also recommends that the Federal Chancellery produce an integrated *Global Report* in the middle of each legislature period. In view of the processes triggered off by UNCED in Rio de Janeiro, the Report should provide information on Federal Government activities concerning global change and sustainable development. Policymaking and research activities in Germany should be analyzed in terms of their environmental, economic and sociocultural impacts within the global network of interrelations. The Council firmly believes that such a report would become an important source of information for the general public in Germany and for foreign institutions, and that it would exert a consolidating and integrating influence on global change activities in the various federal ministries.

The work of the German Parliament's Enquete Commissions has had an integrating effect on German research and government support agencies. An Enquete Commission on "Global Change" could continue the work of the Enquete Commission Protection of People and the Environment, whereby the focus should be placed on implementing the recommendations of the scientific community as advanced, for example, by the German Advisory Council on Global Change.

For some time now, there have been discussions about establishing a *German Academy of Sciences*, similar to those in other countries, which could state its position on issues of national importance with a high degree of independence and authority; were such an academy to be created, the problem of global change would certainly be an important topic for it to consider.

Prospects

In relation to its population, Germany bears a disproportionate responsibility for the causes of global change. Its contribution towards global change research, albeit substantial, must be radically increased. The primary requirement is not so much a major increase in research funding, or the founding of new large-scale research establishments, but the efficient use of data and knowledge already available and the synthesis of that knowledge to solve complex problems. What is also needed are organizational measures to ensure that existing global change research potential is deployed more effectively, and that gaps in the various research fields can be closed by providing a modest level of extra funding.

Transnational networking and integration into international programs at European and global level are crucially important for German global change research. In accord with Germany's role in the world economy, German research should play a leading role in creating and expanding research capacities in the developing countries.

Introduction

A

For the first time in history, human activities are having impacts of planetary scale. The resultant changes in the global environment are reshaping the relationship between humankind and the natural basis on which its existence depends. This transformation process, called *global change*, is occurring at unprecedented speed and involves many risks. It can only be understood if Earth is conceived of holistically as a single system. Global change poses a major challenge for the scientific community, which must describe and explain how the Earth System is altered by human intervention, how these processes are influenced in turn by natural changes in the Earth System, and, finally, whether and to what extent there are ways to control global change.

Anthropogenic climate change is a good illustration of the sheer dimensions of human-induced effects on the global environment. Carbon dioxide emissions from the transport sector in Germany contribute towards rising sea level and the expected disappearance of coral islands 20,000 km away, thus robbing the inhabitants of such islands of their very habitat. Humankind is thus confronted not only with a major ethical dilemma, but also with a complex of difficult research issues and problems which have to be resolved as quickly and as competently as possible. These problems can only be surmounted by *interdisciplinary* and *international* research networks in which, alongside climate modeling and hydrography, for example, other disciplines such as the philosophy of law and cultural anthropology must also play a role.

Global change research poses an enormous challenge for the researchers themselves, for the bodies funding and promoting such research as well as for those who base their decision-making on its findings, and demands integrative capacity, flexibility and imaginative power from scientists, research promoting organizations and users. Innovative guidelines and structures are necessary for handling the complex problems and for developing the requisite problem-solving competence. "Traditional" environmental research has proved so far to be inappropriate for meeting these demands.

In its 1993, 1994 and 1995 Annual Reports, the Council identified and described the core problems associated with global change – on the one hand, the changes in people's natural environment (the "ecosphere"), on the other, the changes within human society itself (the "anthroposphere"). The 1996 Annual Report now focuses on the organization of global change research and examines the conditions for enhancing Germany's contribution.

Suspicions have often been voiced that calls for new research serve only to divert attention from the necessity of environmental action. However, research explicitly geared to *solving specific problems* is indeed relevant for political action, improving the decision-making competence on which any such action depends.

The Council's focus in this year's Annual Report embraces not only the "classical" fields of environmental research within the natural sciences, but also the economic and sociocultural dimensions of global change. The methodological foundations were established in previous Reports through the development of an integrative research approach based on the *analysis of "syndromes"* (WBGU, 1993 and 1995a). This approach permits the operationalization of the networked thinking that is essential for mitigating global change and offers new options for the design and organization of global change research.

When public funds are scarce, clear-cut priorities are required in order to achieve maximum efficiency in the selection and execution of research projects. To this end, the Council has elaborated *relevance criteria* and *integration principles* for global change research. These criteria and principles could also be applied in the current reshaping of Germany's environmental research programs. The Council welcomes the interministerial initiative of the *Bundesministerium für Bildung, Wissenschaft, Forschung und Technologie* (BMBF) (Federal Ministry for Education, Science, Research and Technology) and the *Bundesministerium für Umwelt, Naturschutz und Reaktorsicherheit* (BMU) (Federal Ministry for Environment, Nature Conservation and Reactor Safety) for a new environmental research program to succeed several specialized programs and government support frameworks for global change research in the fields of climate, marine, polar and ozone research.

The analysis of the current status of global change research presented in *Section B* of the Report identifies a number of deficiencies: German research is not integrated sufficiently at the international level; there are weaknesses in cooperation between the various disciplines and in the development of interdisciplinary approaches; organizational networking and coordination needs to be expanded and intensified in both the scientific and administrative dimensions of global change research, and there is not enough communication between the research community, the political sphere and society.

The Report points out ways to eliminate these deficiencies. Building on its systems approach, the Council formulates in *Section C* a new set of guiding principles for shaping environmental research in general, and global change research in particular. These ideas lead in *Section D* to recommendations for research with wider implications for the organization of scientific endeavor in Germany.

Status of Global Change Research, Unexplored Issues

Global change cannot be studied from a purely national perspective, but requires a global approach. The last twenty years have seen the establishment of a wide range of international research programs addressing this increasingly topical issue. This year's Annual Report commences therefore with an overview of these programs, covering not only the research activities of global change programs but also integrative activities (e.g. data management and capacity building) and monitoring programs.

The first global change research programs primarily involved disciplines in the natural sciences, and were geared to the identification, analysis and interpretation of global climate changes. This was the context for the establishment of the World Climate Research Programme (WCRP), which has been studying the global climate system since 1979, and which is aimed at predicting climate changes at global and regional level. The Intergovernmental Panel on Climate Change (IPCC) was set up through the joint initiative of the World Meteorological Organisation (WMO) and the United Nations Environmental Programme (UNEP) to facilitate the translation of research results into political action. The IPCC is an international body of scientists and government representatives that produces reports on global climate changes. The IPCC assessments contain the information on which the Climate Convention bodies base their work. The International Geosphere-Biosphere Programme (IGBP) focuses on the interactions between biota and the environmental spheres they inhabit and have impacts on, namely the atmosphere, hydrosphere and lithosphere.

The central focus of the International Human Dimensions of Global Environmental Change Programme (IHDP) is the way in which human societies both cause and are affected by global change. Also established in 1986, IHDP has recently undergone major restructuring and is now consolidating its activities.

A number of other research programs, foremost among them the Man and the Biosphere (MAB) program operated by UNESCO, reflect the increasing amount of attention that is now devoted to the cru-cial role played by human beings in global change. Various global change research efforts have also been initiated within a European framework. Finally, mention should also be made of the major monitoring programs currently in place, which provide an essential basis for conducting global change research. Special reference is made in the following to the monitoring programs within the Earthwatch activities of the UNEP (*Fig. 1*).

1.1
World Climate Research Programme (WCRP)

The World Climate Research Programme (WCRP), a sub-project of the World Climate Programme (WCP), was set up in 1979 by the ICSU (International Council of Scientific Unions) and WMO. The goal is to determine the extent to which climate and anthropogenic influences on climate can be predicted. This calls for a quantitative understanding of the four primary components of the physical climate system, namely the atmosphere, the oceans, the polar cryosphere and the land surface of the continents.

1.1.1
Organizational Structure of the WCRP

The research priorities of WCRP are established by the Joint Scientific Committee (JSC), comprising representatives of the WMO, the ICSU and, since 1993, the Intergovernmental Oceanographic Commission (IOC) of UNESCO. The main projects proposed by the JSC each have their own secretariat (*Fig. 2*). Germany hosts the secretariat of the Climate Variability and Predictability Programme (CLIVAR), which is jointly funded by Australia, Germany and Japan. The various projects are coordinated at international level by the Joint Planning Staff (JPS) in Geneva. Research planning is the responsibility of the Scientific Steering Groups (SSG), whereby implementation of the projects is dependent on the ef-

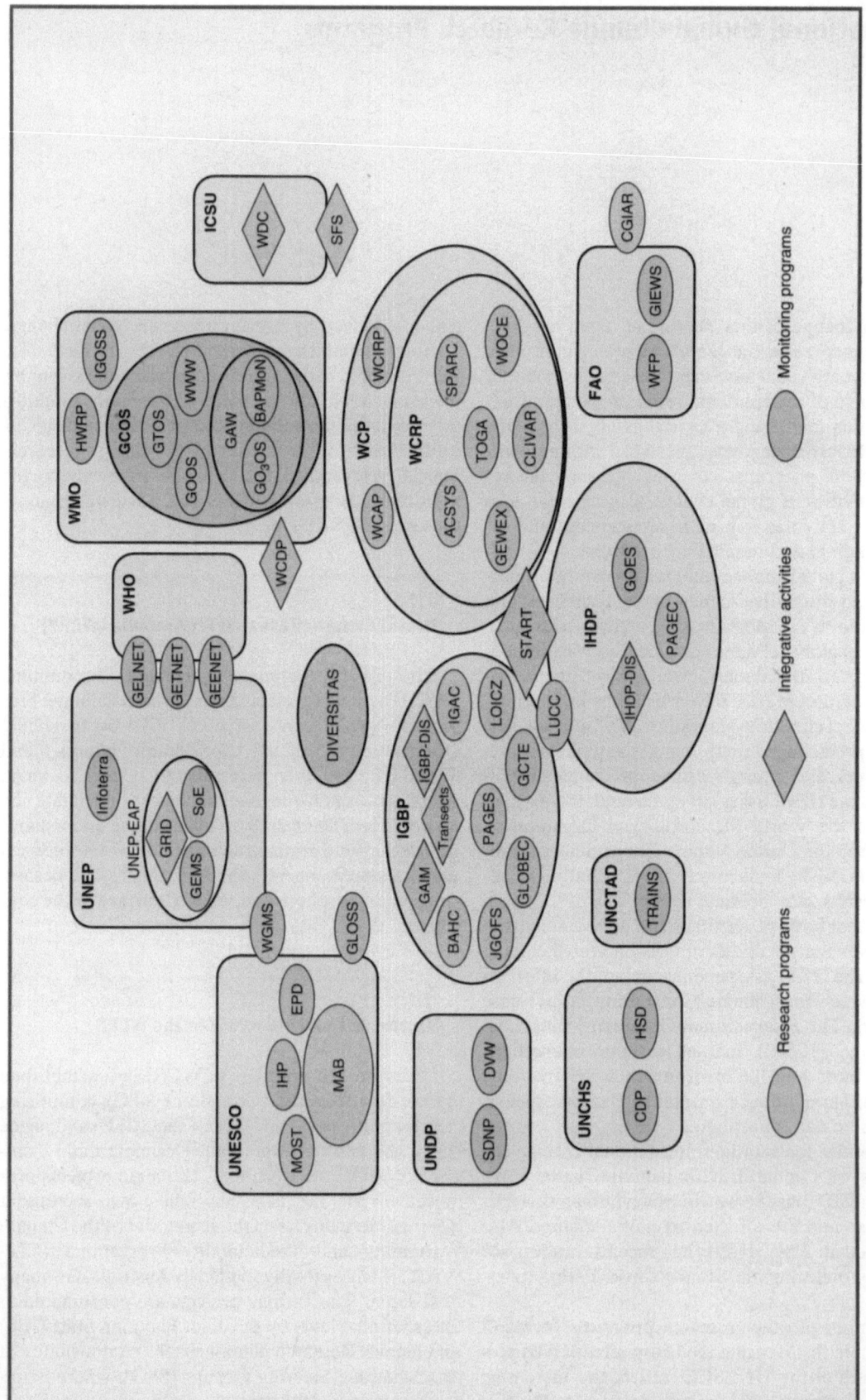

Figure 1
International Global Change Research Programs.
Source: WBGU

BOX 1

Acronyms in Fig. 1

ACSYS: Arctic Climate System Study

BAHC: Biospheric Aspects of the Hydrological Cycle

BAPMoN: Background Air Pollution Monitoring Network

CDP: City Data Programme (in preparation)

CGIAR: Consultative Group on International Agricultural Research

CLIVAR: Climate Variability and Predictability Programme

DIVERSITAS: Program for biodiversity research

DVW: Development Watch

EPD: Environment and Population Education & Information for Development

FAO: Food and Agriculture Organisation

GAIM: Global Analysis, Interpretation and Modelling

GAW: Global Atmosphere Watch

GCTE: Global Change and Terrestrial Ecosystems

GCOS: Global Climate Observing System

GEENET: Global Environmental Epidemiology Network

GELNET: Global Environmental Library Network

GEMS: Global environmental monitoring System

GETNET: Global Environmental Technology Network

GEWEX: Global Energy and Water Cycle Experiment

GIEWS: Global Information Early Warning System

GLOBEC: Global Ocean Ecosystem Dynamics

GLOSS: Global Sea Level Observing System

GOES: Global Omnibus Environmental Survey

GOOS: Global Ocean Observing System (in preparation)

GO_3OS: Global Ozone Oberserving System

GRID: Global Resources Information Database

GTOS: Global Terrestrial Observing System

HSD: Human Settlements Database (in preparation)

HWRP: Hydrology and Water Resources Programme

ICSU: International Council of Scientific Unions

IGAC: International Global Atmospheric Chemistry Project

IGOSS: Integrated Global Ocean Service System

IGBP: International Geosphere Biosphere Programme

IGBP-DIS: IGBP-Data and Information System

IHDP: International Human Dimension of Global Environmental Change Programme

IHDP-DIS: IHDP-Data and Information System

IHP: International Hydrological Programme

Infoterra: International Environmental Information System

IOC: Intergovernmental Oceanic Commission

JGOFS: Joint Global Ocean Flux Study

LOICZ: Land-Ocean Interactions in the Coastal Zone

LUCC: Land-Use and Land-Cover Change

MAB: Man and the Biosphere Programme

MOST: Management of Social Transformations

PAGEC: Perceptions and Assessment of Global Environmental Conditions and Change

PAGES: Past Global Changes

SDNP: Sustainable Development Networking Programme

SFS: Sciences for Food Security (in preparation)

SoE: State of the Environment

SPARC: Stratospheric Processes and their Role in Climate

START: System for Analysis, Research, and Training

TOGA: Tropical Ocean - Global Atmosphere (until 1994)

TRAINS: Trade Analysis and Information System

UNCHS: United Nations Centre for Human Settlements (HABITAT)

UNCTAD: United Nations Conference on Trade and Development

UNDP: United Nations Development Programme

UNEP: United Nations Environment Programme

UNEP-EAP: UNEP Environment Assessment Programme

UNESCO: United Nations Educational, Scientific and Cultural Organization

WCAP: World Climate Assessment Programme

WCDP: World Climate Data Programme

WCIRP: World Climate Impacts and Response Strategies Programme

WCP: World Climate Programme

WCRP: World Climate Research Programme

WDC: World Data Centres

WFP: World Food Programme

WGMS: World Glacier Monitoring Service

WHO: World Health Organization

WMO: World Meteorological Organization

WOCE: World Ocean Circulation Experiment

WWW: World Weather Watch

forts and resources deployed by the individual countries. The WCRP itself has a very restricted budget for providing the organizational infrastructure. *Box 2* lists the main WCRP projects.

1.2
International Geosphere Biosphere Programme (IGBP)

The International Geosphere Biosphere Programme (IGBP), established in 1986 by the ICSU, is dedicated to the investigation of interactive processes and cycles within the geosphere and the biosphere, as well as their feedback effects on global change. The geosphere, i.e. the totality of physical and chemical environment, and the biosphere, the complex of

manifold interrelationships and interdependencies between living beings, including humankind, should be understood as a coupled unity. Multidisciplinary, internationally coordinated projects focus on the natural interactions, increasingly influenced by human activities, between soils, water and air, and their feedbacks on regional and global climate. The aim of IGBP is "to describe and understand the interactive physical, chemical and biological processes that regulate the total Earth system". This overall aim encompasses a very broad spectrum of research activities. Since it is essential in this context to set priorities, IGBP focuses on those processes that affect global change across time scales ranging from decades to centuries that are most sensitive to anthropogenic perturbations and that will most likely lead to practical, predictive capability.

BOX 2

Main Projects Within WCRP

- ACSYS (since 1994): The Arctic Climate System Study focuses on the observation and modeling of the Arctic Ocean, as well as the formation, transport and melting of sea ice. One aim is to discover the mechanisms by which freshwater is exported from the Arctic to the north Atlantic; this will provide insights into changes in deep water formation and the related effects on global thermohaline circulation. Secretariat: Oslo, Norway.
- CLIVAR (since 1995): The Climate Variability and Predictability Programme aims at improving our understanding of natural climate variability, detecting anthropogenic changes and predicting climate on seasonal to centennial time scales. CLIVAR builds on the successful work carried out in the TOGA and WOCE projects, and consists of three main component programs: 1) a study of the seasonal to interannual variability and predictability of the global ocean-atmosphere-land system (GOALS), 2) decadal to centennial climate variability and predictability, with special emphasis on the role of the oceans in the global coupled climate system (DecCen), and 3) modeling and detection of anthropogenic climate change. Secretariat: Hamburg.
- GEWEX (since 1988): The Global Energy and Water Cycle Experiment was set up to study, model and predict the transport and exchange

of radiation, heat and water in the atmosphere and at the earth's surface, and to analyse the effects of climate change on global and regional precipitation regimes. Germany provides special support for the project focusing on water resources in the Baltic Sea catchment. Secretariat: Silver Spring, USA.
- SPARC (since 1993): The investigation of Stratospheric Processes and their Role in Climate aims at identifying how stratospheric processes have feedback impacts on climate. This project extends the focus of WCRP research work to the stratosphere. Germany contributes to SPARC primarily through the ozone research program (OFP), and by operating several stations for measuring ozone and UV radiation. Secretariat: Verrières-le-Buisson, France.
- TOGA (1985-1994): The study of Tropical Ocean - Global Atmosphere interactions generated new insights into the causes of the Southern Oscillation and the related El Niño phenomenon. Germany's involvement in TOGA was mainly in the modeling field.
- WOCE (1990-2002): The World Ocean Circulation Experiment aims to provide the first-ever virtually simultaneous observations of all the world's oceans as a basis for constructing realistic mathematical models of global circulation and heat transport in the oceans. Physical oceanographers in Germany have dedicated a substantial proportion of their resources to the WOCE research effort. Secretariat: Southampton, Great Britain.

Figure 2
Organizational structure
of the WCRP.
Source: WBGU

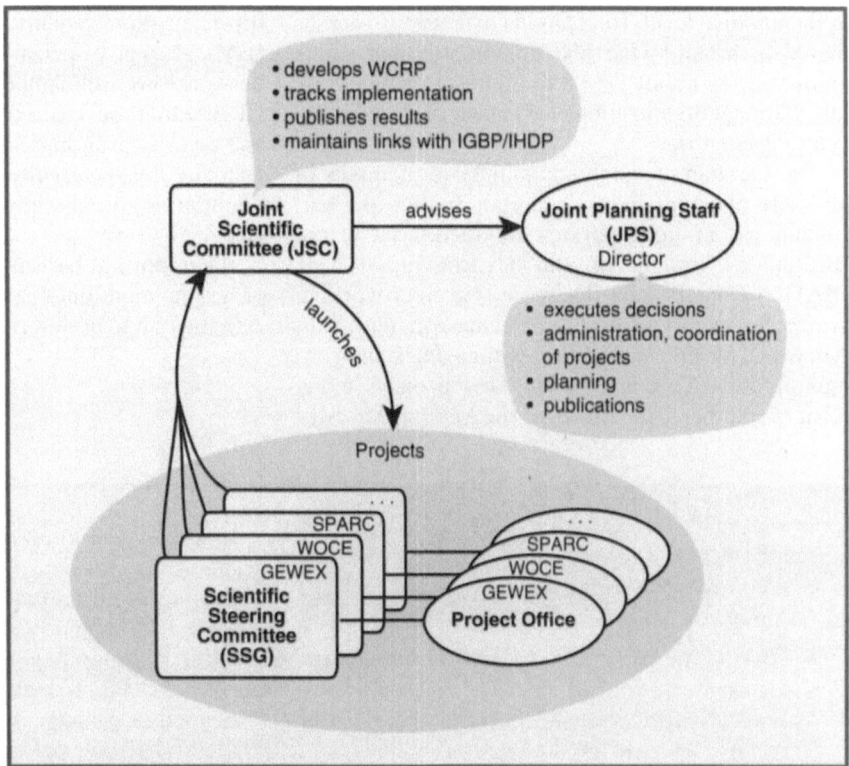

The initial task of IGBP was to define critical gaps in our understanding of global biogeochemical cycles and life support processes. The next step was to formulate a number of "core projects" addressing these gaps. IGBP officially commenced its operations in 1990 by adopting a formal organizational structure, including the IGBP Scientific Committee (SC-IGBP). Provisional planning extends beyond the year 2000 (IGBP, 1994). A Core Project Office (CPO) was set up for each core project with responsibility for coordinating the respective research activities (*Box 3*).

In addition, IGBP has three over-arching, integrative activities (*Framework Activities*), serving all core projects – centered on modeling (GAIM), data acquisition and data management (IGBP-DIS) and the promotion of IGBP in developing countries (START). These activities are also geared to establishing closer links between IGBP and other national or international activities in the field of global change research.

1.2.1
The IGBP Organizational Structure

IGBP, like WCRP, has evolved a complex organizational structure typical for major international programs; this structure is mirrored at national level

(*Fig. 3*). The Scientific Committee (SC-IGBP), which consists of independent scientists and the Chairs of the core projects, develops the program, defines priorities and guides its implementation, including publication of results. It takes decisions regarding new core projects. The IGBP Scientific Advisory Council (SAC) consists of representatives of the National IGBP Committees and ICSU member bodies. Its role is to advise on the scientific content of the program, assess its results, and make general policy recommendations. The IGBP Secretariat in Stockholm is responsible for implementing the decisions of the SC-IGBP. More than 60 National Committees act as the formal interface between researchers in participating countries and the international structure. The National Committees play a greater role in developing and newly-industrializing countries than in countries with a firmly established national scientific structure. Each of the core projects and framework activities has its own, largely autonomous system of steering committees and secretariats that in many cases operate in almost complete independence of the IGBP framework. The German National Committee, the Secretariat of which is located at the Potsdam Institute for Climate Impact Research (PIK), coordinates contacts between the German working groups for the separate core projects.

The research activities conducted within the framework of IGBP are financed almost exclusively at the national level. Total annual expenditure is estimated at around US$ 800 million. Limited funds amounting to about US$ 3 million are provided by international organizations to finance projects in developing countries.

The German research community participates in all IGBP planning and coordination bodies to a reasonable extent, and operates the Secretariat of the Biospheric Aspects of the Hydrological Cycle (BAHC) Core Project. There are German working groups for all core projects and framework activities. Almost all the German institutes engaged in environmental research are involved in IGBP projects in one form or another. The efforts of the Federal Ministry for Education, Science, Research and Technology (BMBF) to give special support to national research networks are well-suited to IGBP's goal of integrated research. Several collaborative research centers and priority programmes supported by the *Deutsche Forschungsgemeinschaft* (DFG) (German Research Foundation) are directly involved in IGBP core projects.

An important basis for the global change research programs mentioned so far are parallel monitoring activities, an overview of which is provided in *Box 4.*

BOX 3

IGBP Core Projects

- BAHC: The Biospheric Aspects of the Hydrological Cycle core project studies the interactions of vegetation with physical processes of the hydrological cycle. Secretariat: Potsdam, Germany.
- GCTE: The Global Change and Terrestrial Ecosystems project looks at the effects of global change on terrestrial ecosystems. Secretariat: Lynham, Australia.
- GLOBEC: The objective of the Global Ocean Ecosystem Dynamics project is to predict potential reactions of the ocean system to global change phenomena. Secretariat: not yet decided.
- IGAC: The central research questions addressed by the International Global Atmospheric Chemistry project concern the processes which regulate the chemistry of the global atmosphere, the role of the biosphere in global trace gas cycles, and aerosols. Secretariat: Cambridge, USA.
- JGOFS: Ocean biogeochemical cycles, and the interactions of these cycles with climate, are the priority issues dealt with in the Joint Global Ocean Flux Study. Secretariat: Bergen, Norway.
- LOICZ: The impact of land-use changes, rising sea level and climate change on coastal ecosystems are the subject of the Land-Ocean Interactions in the Coastal Zone project. Secretariat: Texel, Netherlands.
- LUCC: Changes in land use and land cover, their future development, and the interactions between land use, land cover and climate change are investigated in the Land-Use and Land-Cover Change project. LUCC is the only project to be jointly established by IGBP and IHDP. Secretariat: not yet decided.
- PAGES: "What significant climatic and environmental changes occurred in the past, and what were their causes?" These are the central questions addressed by the Past Global Changes project. Secretariat: Berne, Switzerland.

INTEGRATIVE ACTIVITIES CONDUCTED BY THE IGBP

- GAIM: The Global Analysis, Interpretation and Modelling Task Force is responsible for promoting the development, evaluation and application of comprehensive predictive models of the global biogeochemical system, and subsequently linking such models to those of the physical climate system. Secretariat: Durham, NH, USA.
- IGBP-DIS: The IGBP Data and Information System assists the core projects in meeting their data acquisition and data management needs and facilitates collaboration with space agencies and international data-producing bodies. Secretariat: Toulouse, France.
- START: The purpose of the Global Change System for Analysis, Research and Training is to promote regional capacity building in global change research, especially in developing countries. The aim of this joint integrative activity sponsored by IGBP, WCRP and IHDP is to establish efficient networks linking regionally based research and analysis projects. Secretariat: Washington D.C., USA.

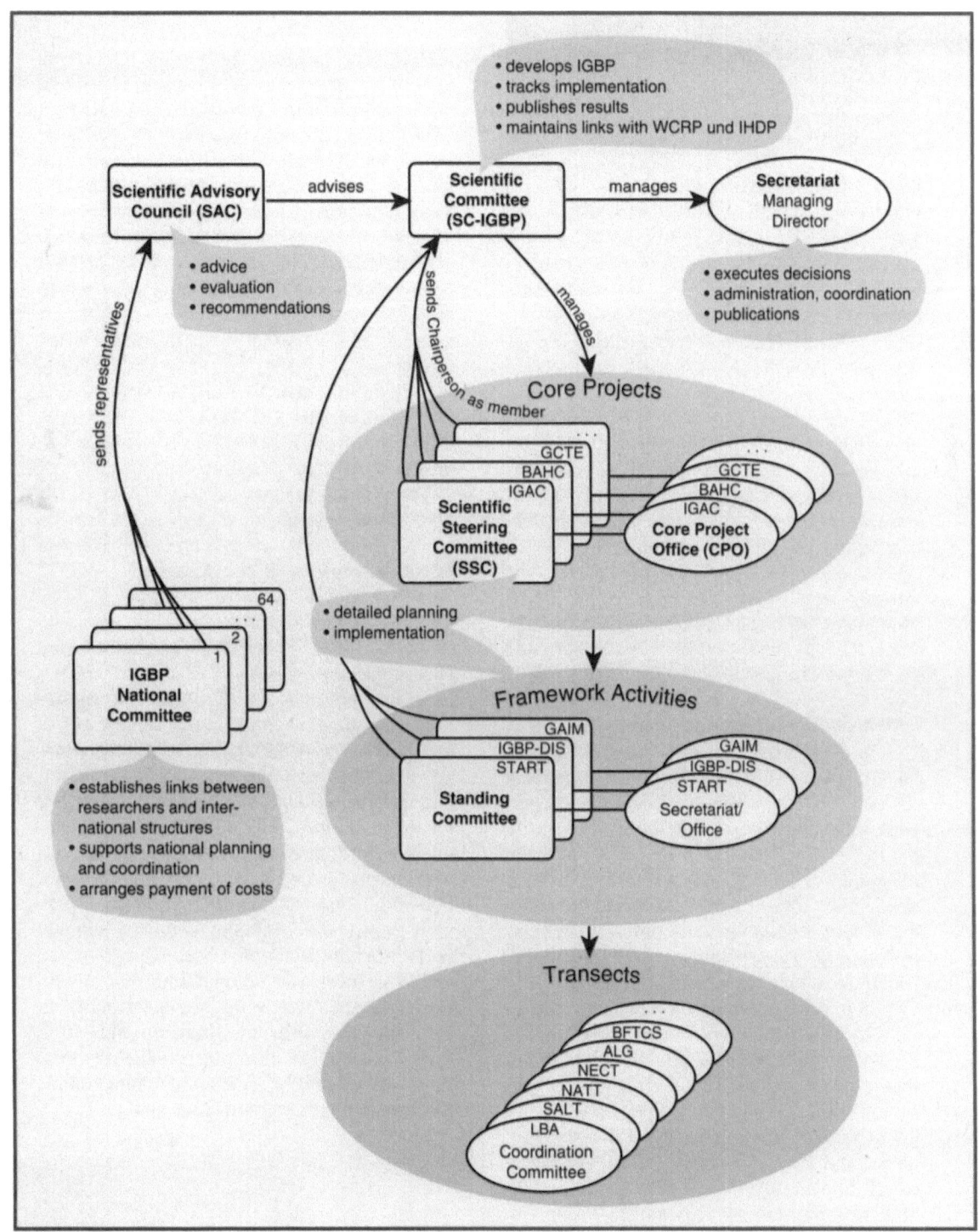

Figure 3
Organizational structure of the IGBP.
Acronyms: BFTCS *Boreal Forest Transect Case Study*, ALG *Alaskan Latitudinal Gradient*, NECT *North East Chinese Transect*, NATT *North Australian Tropical Transect*, SALT *Savanna on the Long-Term*, LBA *Large Scale Biosphere-Atmosphere Experiment in Amazonia*. Further acronyms see *Box 1*.
Source: WBGU

BOX 4

International Global Change Monitoring
Programs

- GCOS: The objectives of the Global Climate Observing System, jointly established by WMO, IOC and ICSU, are to provide and ensure the continuity of systematic and comprehensive observation of the global climate system, to detect climate changes and to monitor the impacts that such changes have on terrestrial ecosystems. GCOS aims to integrate existing observation systems as far as possible and, where necessary, to develop new, additional systems of this kind. Moreover, the project aims at ensuring the continuity of measurements, which in turn requires regular support from the participant countries. GCOS also encompasses the Global Atmosphere Watch (GAW) observation system. GAW includes the Global Ozone Observing System, the Background Air Pollution Monitoring Network, the Integrated Global Ocean Services System and the Global Sea Level Observing System. Data from the World Weather Watch Programme (WWW) are also integrated into the system.
- GOOS: A network of manned and automatic meteorological stations spans the continents of the world, but there is no such observation network for the world ocean. In cooperation with the WMO and the ICSU, the IOC has been planning a Global Ocean Observing System since 1993. This involves technical developments and installations amounting to billions of dollars. It is also planned to integrate global communication networks and data centers into GOOS in order to attain the objectives set for GCOS. Another GOOS module involves the assessment and prediction of the "health" of the ocean. EUROGOOS is a regional organization within GOOS that was set up to ensure intensive participation of the European nations (and hence German industry and research bodies) in the development of marine observation systems specifically geared to European interests.

- GTOS: The Global Terrestrial Observing System is currently in the planning phase and will be closely linked to GCOS and GOOS.
- UNEP programs: All UN environmental monitoring activities have been concentrated since 1972 in Earthwatch. The purpose of Earthwatch is to report, coordinate, harmonize and integrate observation data in order to anticipate environmental changes and to provide policymakers with the information on which to base decision-making on environmental measures. UNEP participates in Earthwatch through its monitoring programs GEMS (Global Environmental Monitoring System) and the sub-programs GEMS/Air, GEMS/Water, GEMS/Food and GEMS/HEALS (Human Exposure Assessment Location Project).

  The management and dissemination of the observation data acquired through the various GEMS programs is ensured by GRID (Global Resource Information Database), IRPTC (International Register of Potentially Toxic Chemicals) and INFOTERRA. The responsibility of GRID is to collect and consolidate geographically explicit data from the GEMS programs operated by international institutions and organizations. Geographical Information Systems (GIS) enable the interrelationships between regional data sets to be identified and visualized. IRPTC is a global network for the acquisition and provision of data relating to hazardous chemicals. INFOTERRA is a mechanism for exchanging data and information within and between states. A major priority is to provide developing countries with access to such information.
- GIEWS: The FAO's Global Information and Early Warning System was set up as an instrument for monitoring and predicting the world food situation. The objective is to identify food supply crises at an early stage to enable timely and targeted interventions to be made.

1.2.2
Evaluation of IGBP from the International Perspective

In 1994/95, at the behest of the sponsor organization ICSU and the International Group of Funding Agencies for Global Change Research (IGFA), an international group evaluated the development of IGBP. The overall assessment was positive: the various components (Core Projects and Framework Activities) generate useful output for the scientific community and society in general, while the IGBP's coordinating activities provide the national and international projects with internal coherence and external presence. The predominantly unbureaucratic way in which IGBP and its core projects are organized is adequate for a program that lives from the initiative of individual scientists and national planning groups. The report explicitly warns against greater formalization of IGBP and the indefinite continuation of its various activities, and against enlarging its very efficient Secretariat. The latter, however, should collaborate and communicate more intensively with the Secretariats of the core projects.

The first International IGBP Congress took place in April 1996 in Bad Münstereifel, Germany, following completion of the implementation phase of most core projects and the first evaluation of the IGBP. An outstanding example for the new integrative and interdisciplinary activities within IGBP, the *Transect Activities*, were debated at the conference. Within the framework of these activities, IGBP has been conducting regional studies in all major climatic and vegetation zones over the last 10-15 years. The intention is to develop these into a new integrative framework cross-cutting the individual core projects. Such "interprojects" are expected to establish stronger links between the needs of fundamental research and issues of political relevance, demonstrate the linkages between local processes and global mechanisms, and be aligned better with the funding structures of the agencies. In future, these transect activities will require a high level of input from the social sciences.

Increases in efficiency can be expected for IGBP when the major international programs strengthen their contacts: those to WCRP are already good, while collaboration with DIVERSITAS (*see Box 5*) and IHDP (*see Section B 1.3*) is still very weak. Recommendations that IGBP become more involved than hitherto in the development and activities of the international conventions on climate, biodiversity, desertification, etc. point in the same direction. IGBP has recently started to move away from a purely natural-scientific focus and is now integrating the socioeconomic aspects of environmental change. One example is the joint IGBP/IDHP core project Land-Use and Land-Cover Change (LUCC) (*see Section C 1.3*).

1.3
The International Human Dimensions of Global Environmental Change Programme (IHDP)

The goal of the International Human Dimensions of Global Environmental Change Programme (IHDP, previously HDP, HDGC, HDGEC) is to develop and promote social science research initiatives of primary relevance for understanding the role of human activities in bringing about global environmental changes and the consequences of these changes for human beings and human societies.

IHDP was initiated by the International Social Science Council (ISSC) in 1990 after a four-year exploratory phase, but since then has undergone little further development compared to IGBP. Since February 1996, it has been co-sponsored by ICSU. IHDP sees itself as the social science parallel or complement to the natural science programs IGBP and WCRP. Being a framework program, IHDP aims to promote the exchange and networking of social science research on global change.

Within its broad framework program, specific work plans have been elaborated only for the Perception and Assessment of Global Environmental Change (PAGEC) core project and the Global Omnibus Environmental Survey (GOES) monitoring program. The objective of the latter is to develop a methodological instrument for worldwide monitoring of environmental knowledge, attitudes and behavior. Realizing the major importance of changes in land use and land cover, IGBP and IHDP have established a joint core project on Land-Use and Land-Cover Change (LUCC). Other research topics discussed so far by IHDP are Industrial Transformation, Energy Use, Demographic and Social Dimensions of Resource Use (DSDRU), Environmental Security, Trade and Environment, Institutions and Health.

IHDP collaborates closely with the Consortium for International Earth Science Information Network (CIESIN) on the electronic archiving and worldwide sharing of social science data, one aim being the establishment of the IHDP Data and Information System (IHDP-DIS) as a framework activity in the service of the program.

In addition to the workshops organized by the respective working groups and the meetings of the IHDP Committees, scientific symposia are held about every two years, at which the overall thrust of the program and the location of its activities are discussed. IHDP operates a Secretariat in Geneva;

funding is still obtained for individual projects or on the basis of grants. Efforts are being made to arrange a basic level of funding, for example through the research funding agencies at national level.

Following the Third IHDP Scientific Symposium in Geneva in 1995 and the involvement of ICSU as co-sponsor, the program is now in a phase of restructuring. At the first session of the new Scientific Committee in May 1996, it was agreed to focus IHDP activities on a small number of core projects, but to intensify the research work they perform. Efforts are also being made to establish National HDP Committees. Existing research activities at national level will also be networked with IHDP activities.

1.4
The UNESCO "Man and the Biosphere" Program

In addition to WCRP, IGBP and IHDP, there are a large number of other international programs relating to the problems of global change, some of which include intensive participation of the part of the German research community. One of the most important of these is the UNESCO program "Man and the Biosphere".

1.4.1
Research Priorities and Goals

Man and the Biosphere (MAB) was launched in 1971 by the UNESCO's General Conference. The mission of the MAB program is to develop and improve, at international level, the scientific basis for environmentally sustainable use of the biosphere and in this way to help towards solving global environment and development problems. If this goal is to be attained, then human activities must be included in the analysis. This extended ecosystem approach addresses both environmental and societal aspects (cultural, social, economic, etc.). A special focus of MAB is to develop basic concepts and models for sustainable use of the biosphere. These models are developed, tested and implemented at selected sites called biosphere reserves. The priority focal points of the program were redefined in 1993 in response to the UNCED Conference in Rio de Janeiro (Erdmann and Nauber, 1995):

- Protection of biodiversity and ecological processes.
- Development of strategies for sustainable use.
- Promotion of communication networks and environmental education.
- Capacity building in the field of training.

- Establishment and operation of a global environmental monitoring system.

A cardinal function of the MAB program is to promote and support the designation of biosphere reserves worldwide. By 1995, the biosphere reserve network had grown to 328 reserves in 82 countries. The German MAB National Committee is especially committed to the launching of a Biosphere Reserve Integrated Monitoring program (BRIM). Research projects in some biosphere reserves have been collecting environmental monitoring data for several years now. By collating and analyzing existing data sets and by systematically monitoring other parameters, MAB will attempt to describe the current state of the European environment and predict its future trajectory. Social aspects will also be taken into account, in accordance with the basic philosophy underpinning the MAB program. Environmental monitoring by MAB should be coordinated with the respective national activities in this field.

1.4.2
Organizational Structure and International Cooperation

The MAB Secretariat is located at UNESCO headquarters in Paris. An International Coordinating Council (ICC) is responsible for international organization, planning and coordination of the MAB program (*Fig. 4*). The ICC is elected for a term of four years at UNESCO's General Conference. In addition to the workshops conducted by the various working groups, ICC conferences are held about every two years.

In response to the global dimension of environmental changes, the MAB program has been international in scope from the very outset. Special emphasis is placed on the integration of developing countries into the program. The implementation and organization of the MAB program is the responsibility of the MAB Secretariat, which is comprised of one representative from each of the UN regions – Africa, Arabia, Asia/Australia, Latin America, Western Europe and Eastern Europe. National MAB Committees are appointed by the governments of the UNESCO Member States. These National Committees are primarily responsible for specifying the focal points of national activities within the international programs.

The Office of the National MAB Committee in Germany is attached to the Federal Agency for Nature Conservation (BfN) in Bonn. The focal points of the German contribution are the designation of additional biosphere reserves and the development of ec-

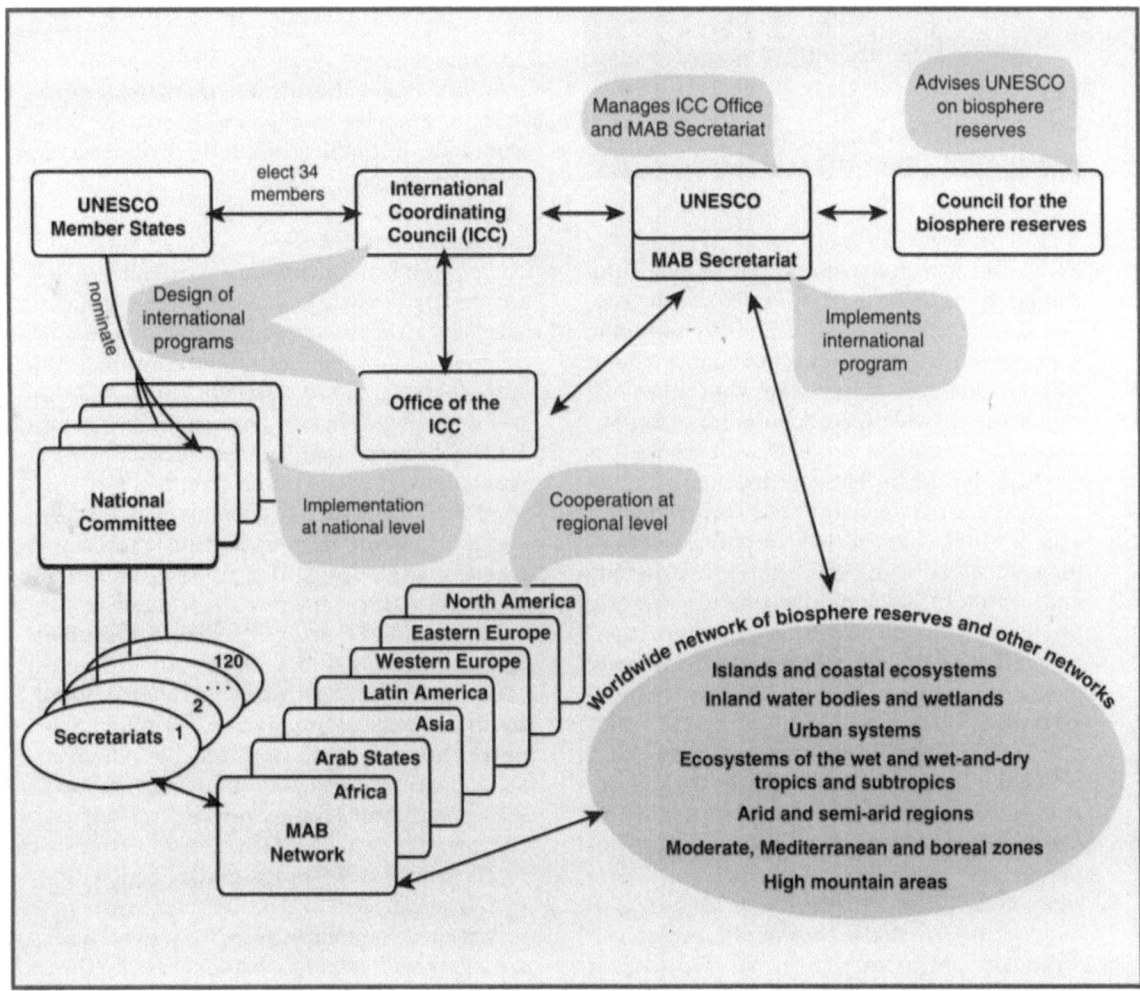

Figure 4
Organizational structure of MAB.
Source: WBGU

osystem research centers and related research programs funded by the BMBF, the BMI, the BMU, the DFG and the relevant *Bundesländer*. Biosphere reserves will continue to be a central element in Germany's contribution to MAB. The key research issues covered by the next medium-term national program (following that for the 1996-2001 period) will be Sustainable Development, Environmentally Responsible Action, Biodiversity and Land-Use Changes.

Biosphere reserves play an important role in international cooperation networks. Projects relating to biosphere reserves can be integrated into development cooperation through the National MAB Committees of the individual developing countries. Germany's contribution to international cooperation also involves the transfer of trustee funds to UNESCO. Worldwide collaboration is also organized

through the National MAB Committees, the MAB Secretariat in Paris and through regional networks (at continental and sub-continental level). Regional MAB networks currently exist in Europe and North America (EUROMAB), are under construction in Central and Latin America, and are being planned in West Africa. Collaboration is also possible among the National MAB Committees or on a purely bilateral basis.

Other important international research programs that address global change issues are described in *Box 5*.

UN PROGRAMS
- EPD: The Environment and Population Education & Information for Development program operated by UNESCO aims to contribute towards solving global development problems through intensive networking with other UN organizations, with special reference to the various world conferences that have been held so far (e.g. the Earth Summit, the International Conference on Population and Development). The priority issues include sustainable development, global and local perspectives, dignity and worth of the human individual, accepting the diversity of life styles, and global partnership. EDP seeks first and foremost to integrate and to bring actors together at the international level.
- GEF: The Global Environmental Facility is a UN body administered primarily by UNDP and the World Bank. A range of individual projects is supported with World Bank funding totalling US$ 200 million. Regional projects are supported in four areas where effects are global in scale (transboundary waters, climate, biodiversity and ozone depletion). Research is not the primary objective guiding the provision of support, but an important means for achieving the goals of the GEF.
- IHP and OHP: The International Hydrological Programme of the UNESCO and the Operative Hydrological Programme of the WMO are collaborating closely on the global inventorying of freshwater reserves, research into the hydrological cycle, and extreme hydrological events. IHP focuses primarily on scientific issues, whereas OHP is concerned with operative aspects from the development of measurement networks to the generation of predictive capacities.
- MOST: The UNESCO research program Management of Social Transformations was launched in 1993 to investigate the social dimensions of rapid global transformation. The initial conception was that sustainable development depends directly on the eradication or mitigation of social and economic problems. The three main themes covered by MOST are:

cities as arenas of accelerated social transformations, multicultural and multiethnic societies, and coping locally and regionally with economic, technological and environmental transformations.

EUROPEAN PROGRAMS
- EU programs in the field of environmental research: The Fourth Framework Programme in the Field of Research and Technological Development and Demonstration (1994 to 1998, with a budget of approximately US$ 12 billion) promotes global change research in the area of RTD (Research and Technological Development) and Demonstration Programmes. The framework program is implemented through specific programs developed within each activity and provides for participation by enterprises, universities and research organizations within the EU. Germany should exert its influence to ensure that new funding for EU programs is distributed evenly between natural scientific/technological and socioeconomic aspects. This objective can only be attained through advance collaboration with EU partners, for example by inviting interested EU Member States to a workshop on this issue.

The specific RTD programs developed within the various activities of the framework program pursue a multidisciplinary approach and are organized on a transnational basis. Three RTD programs in particular involve global change research to a greater or lesser degree:
  - Environment and climate: This program will receive a total of US$ 601 million over the 1994-1998 period. It covers four research areas: 1) the natural environment, environmental quality and global change (with special emphasis on climate change), 2) environmental technologies, 3) space techniques applied to environmental monitoring and research and 4) human dimensions of environmental change. The latter field has been allocated 6.7% of the total funding for the program.
  - Marine science and technology (MAST III): This program has an allocated budget of US$ 258 million and comprises four areas of activity: 1) Marine sciences, 2) Strategic marine research, 3) marine technology and 4) supporting initiatives. The emphasis of strategic marine research is on changes in the coastal zone, and on the resultant social

▶

and economic impacts. However, socioeconomic aspects are not included in the list of research tasks specified for this activity.

– Agriculture and fisheries (FAIR): Research in this program, which receives a budget of US$ 686 million, is conducted in the following areas: 1) agriculture, 2) fisheries and aquaculture (including marine ecosystems and socioeconomic aspects), 3) food technologies, 4) production and processing of biomass, 5) forestry (sustainable development) and 6) rural development (especially socioeconomic aspects).

There is also potential for global change research projects within the specific programs on information and communications technologies, life sciences and technologies, energy, transport, and in the field of Targeted Socioeconomic Research (TSER). Although the socioeconomic aspects of research are emphasized in the general objectives for RTD programs, the allocation of EU funding for environmental research displays a strong bias towards technology and the natural sciences.

The European Network for Research in Global Change (ENRICH) was set up to improve the knowledge base for the development of EU policy options by acting as a clearinghouse for information and cooperation in European global change research. A fundamental principle is to link EU programs with national and international research programs such as IGBP, WCRP and IHDP.

• ESF: The European Science Foundation, established in 1974, has a membership of 59 scientific organizations from 21 countries. The ESF conceives itself as a forum for European research, the long-term development of which it aims to support. The following are some of the programs and projects dealing with global change issues:

– APE (1995-1999): The Airborne Polar Experiment focuses on the stratosphere and trace gases which affect climate.

– EISMINT (since 1993): Outputs from mathematical models are indispensable for improving our understanding of the role of ice sheets in the global climate system. The European Ice Sheet Modelling Initiative was initiated by the joint EU/ESF Committee on Ocean and Polar Sciences (ECOPS).

– EPICA (1996-2000): The Antarctic is one of the world regions containing an enormous wealth of data "archives" on the Earth's ev-

olution. The European Project for Ice Coring in Antartica will be conducting glaciological studies in the region, and involves a major logistical effort.

– GISDATA (1993-1996): Geographical Information Systems: Data Integration and Data Base Design is primarily geared to the coordination and networking of data acquisition, data management and data exchange at European level. The data covers socioeconomic and environmental aspects to equal degrees.

– GRIP (1989-1995): The Greenland Ice Core Project collects long-term data series relating to Earth's evolution from the Greenland ice cores. The aim is to cover a time scale in the order of 500,000 years.

– TERM (since 1995): Tackling Environmental Resource Management is a program for interdisciplinary integration at European level of national programs and projects in the fields of social and behavioral science. The focal points of its activities are: patterns of consumption and production, environmental management under conditions of uncertainty, international cooperation in the management and mitigation of environment/development problems, environmental perception, and communication.

• EERO: The European Environmental Research Organisation is a select group of leading European environment researchers. It acts both as a focus for innovative and interdisciplinary environmental research in Europe and as a network for the interdisciplinary training of young scientists. It provides targeted support for innovative and pioneering interdisciplinary research approaches. The strategic objectives of EERO are achieved through the international post-doctoral fellowship program for supporting outstanding young scientists, short-term fellowships for study visits abroad, the initiation of European environmental research networks, workshops and advanced laboratory courses on current environmental topics, and the preparation of assessments and expertises. Although Germany played a major role in launching EERO (start-up financing provided by the Volkswagen Foundation), it has unfortunately failed to commit itself to regular financial contributions.

• EUREKA: EUREKA is a Europe-wide framework for promoting cross-border cooperation between European enterprises and research establishments in research and technology for the civilian world market.

Environmental research and technologies are one of the principal fields covered by EU-REKA projects, examples being EU 7 EURO-TRAC (Transport and Transformation of Environmentally Relevant Trace Constituents in the Troposphere over Europe) and the follow-up program EUROTRAC II, EU 37 EURO-MAR (umbrella project on the application of advanced marine surveillance technologies) and EU 330 EUROENVIRON (umbrella project comprising activities relating to the ecology of inland water bodies, soil remediation, waste disposal and contaminated sites).

OTHER PROGRAMS

- DIVERSITAS: Initiated by the International Union of Biological Sciences (IUBS), the Scientific Committee on Problems of the Environment (SCOPE) and UNESCO, this program serves to promote and coordinate biodiversity research with the goal of providing information and developing predictive models on the status and sustainable use of biodiversity resources (*see Section B 3.4*).

- IDNDR: The International Decade for Natural Disaster Reduction (1990-2000) is not actually a research program, but aims at helping scientific establishments, development agencies and humanitarian organizations to integrate their activities in the field of disaster reduction. IDNDR was created by the UN General Assembly. The German IDNDR Committee was established in 1989 and is committed to improving precautionary measures, informing the public, initiating scientific programs and setting up operative programs in developing countries.

During the second half of the 1980s, national global change research programs were established by the world's leading industrialized nations. A direct comparison of these programs is fraught with difficulties — research funding in the individual countries is organized too differently, and there are too many differences in focal points and methodologies. The comparison of budgets is barely feasible, considering that the Global Change Research Program in the USA includes very high sums for NASA satellite programs, which are used only partially for global change research. *Tab. 1* contains the global change programs of six selected OECD countries. In addition to basic data on establishment, organizational structure, budget, etc., the table details the main features of the respective research agendas, as well as a comparison and evaluation of the principal research foci in the natural and social sciences.

Germany is not included in this table; firstly, because there is no specific research program on global change, thus would therefore result in basic data and evaluations of a very heterogeneous nature. Secondly, the chapter which follows is an attempt to survey the entire spectrum of global change research in Germany.

Table 1
National Global Change Research Programs.
Sources: CGCP, 1993; EA, 1994; Gray, 1995; Henderson, personal communication, 1996; IACGEC, 1993 and 1996;
Joußen, 1995; Karger, 1992; McLeod, 1995; NRP Programme Office, 1994; RIVM, 1993; RMNO, 1996;

	Great Britain	Japan	Canada
Name of program	UK Global Environmental Change Research Framework	Global Environment Research Program (GERP)	Canadian Global Change Program (CGCP)
Established	in 1990 by the National Environment Research Council (NERC), an advisory body reporting to the Prime Minister	in 1990 by the Council of Ministers for Global Environmental Conservation	1985 by the Royal Society of Canada (independent, non-governmental organization)
Program management	• Inter-Agency Committee on Global Environmental Change (IACGEC): Expert Panel • UK GER Office, which also comprises the Secretariat for the IACGEC	• Japan Environment Agency (JEA) • Center for Global Environmental Research (CGER)	• CGCP Board of Directors • CGCP Research/Policy Committee • CGCP Communications Committee • Secretariat
Institutions involved	Meteorological Office; Overseas Development Administration (ODA); 4 government departments; 5 research councils (including Natural Environment Research Council, NERC and Economic and Social Research Council, ESRC); Ministry of Agriculture, Fisheries and Food (MAFF); Forestry Commission; Environment Agency (EA), etc.	12 ministries; 39 national institutes; Japan Environment Agency (JEA); Science and Technology Agency (STA); Hokkaido Development Agency (HDA); Economic Planning Agency (EPA)	Department Environment Canada; provincial governments; research councils; key federal science departments; private sector (industry); university community; non-governmental organizations; foundations
Budget	The total research expenditures in the global change field amount to approximately US$ 273 million p.a.	US$ 26 million in 1994	Approx. US$ 0.7 million in FY 1995-1996. Total Canadian expenditures in the global change area exceed US$ 36 million p.a.
Publications	"The Globe" newsletter Two directories of GER programs Summary of UK GER: Database		Various series of reports "Delta" newsletter "Changes" bulletin Biannual overview of global change research in Canada

Table 1 (contd.)

Sources (contd.): SGCR and NSTC, 1996; SPP, 1994 and 1995; UK GER Office, 1996

	Netherlands	Switzerland	USA
Name of program	Dutch National Research Programme on Global Air Pollution and Climate Change (NRP)	Swiss Priority Programme Environmental Technology and Environmental Research (SPPE)	U.S. Global Change Research Program (USGCRP)
Established	In 1990 by the Dutch government	In 1991 by the Swiss Federal Parliament. Funding body is the Swiss National Science Foundation	In 1989/90 by the U.S. Government (Global Change Research Act)
Program management	• National Institute for Public Health and Environment (RIVM) and the Royal Netherlands Meteorological Institute (KNMJ) • Steering group • NRP Program Office • Program committee • Program groups of the individual research fields • Evaluation by consulting firms after the 1st phase (1990-94)	• Expert panel (for periodic evaluation of the program, inter alia) • Program Management and Secretariat	• National Science and Technology Council (NSTC) • Commission on Environment and Natural Resources (CENR) • Subcommittee on GC Research (SGCR) • Executive Committee of USGCRP • Coordination Office of the U.S. Global Change Research Program
Institutions involved	12 research institutes; 6 non-governmental institutions (e.g. International Soil Reference and Information Centre, ISRIC); universities	5 universities; 2 technical colleges; Swiss Materials Testing and Research Institute; Paul Scherrer Institute	8 government departments; Environmental Protection Agency (EPA); National Aeronautics and Space Administration (NASA); National Science Foundation (NSF); Smithsonian Institution (SI); Tennessee Valley Authority (TVA), etc. run various agency research programs
Budget	Approx. US$ 45 million for the 2nd phase	The US$ 29.6 million for the 1st period will be cut to 60% p.a. in real terms in the 2nd period	Approx. US$ 1.8 billion in 1996
Publications	"Change" newsletter Internal newsletter RIVM Reports	"Panorama" newsletter	Annual report "Our Changing Planet"

Table 1 (contd.)

	Great Britain	Japan	Canada
Supporting programs and activities	The Global Environment Network for Information Exchange in the UK (GENIE Project) was set up as a network for the collection and dissemination of global change data in the natural and social sciences.	- The Eco-Frontier Fellowship Program was established in 1995 to promote exchange between researchers at the international level - Japan holds the Interim Secretariat of the Asia-Pacific Network for Global Change (APN), whose aim is to further global change research in this region	- Research Panels on special issues (e.g. Fisheries Panel) - Environmental Education project, aims at transmitting global change issues to the educational community - The CGCP acts as a national correspondent for the Inter-American Institute for Global Change Research (IAI) and as national committee for some international programs (e.g. IGBP)
Focal points	Three groups of priority research topics are defined based on the "GER triad" framework comprising human, physico-chemical and biological systems: 1. Underpinning research (primarily unidisciplinary basic research on one of the three systems defined in the GER triad) 2. Interactive research (multidisciplinary research projects, each involving two systems, e.g. impacts on health, agriculture, forestry; past global changes) 3. Systemic research (multidisciplinary research, in which the complete GER triad is involved, e.g. climate studies; land use and water resources; biogeochemical cycles)	Three types of research are carried out in nine research areas and a total of 39 core projects: 1. Impacts of human activities, research on the impacts of global environmental change on human health and ecosystems 2. Research on policy planning 3. Other research Research under this program is structured according to certain categories (core research, integrated research, etc.)	The CGCP will deal with the following issues over the next five years: 1. Global Atmospheric Changes 2. Biodiversity Issues 3. Health and Global Change 4. Other Global Change Initiatives (e.g. Global Environmental Change and Human Security; Environmental Assessment and Global Change, Canadian Fisheries and Global Change) Major activities of the program are summarized under three business lines: - Research assessment and policy options - Research planning and collaboration - Information management and communications

Table 1 (contd.)

	Netherlands	Switzerland	USA
Supporting programs and activities	Development of a program project entitled Development of Policy Options aimed at eliminating shortcomings in the area of know-how transfer.	Discussion fora, which along with the projects will enable a scientific exchange of views and discussion of specific issues (particularly fundamental and methodological issues).	- The Global Change Data and Information System (GCDIS), serves as an information source in support of the program - The U.S. Global Change Research and Information Office (GCRIO) serves to provide information to the interested public
Focal points	In the 2nd phase (1995-2000), the following research fields will be covered in more than 140 projects: 1. Behavior of the climate system as a whole 2. Vulnerablility of natural and societal systems 3. Societal causes and solutions 4. Assessment (synthesis and evaluation of knowledge; dialogue researchers - government - society)	Five "integrated projects" on priority areas have been defined for the 2nd period, 1996-1999; these consist of a certain number of closely linked sub-projects: 1. Climate in the Alpine region 2. Biodiversity 3. Sustainable development in economy and society 4. Sustainable use of soils 5. Waste treatment In addition, project groups will be set up to cover Environmental Problems in Developing Countries.	Four major scientific challenges of the program are defined: 1. Seasonal to interannual climate fluctuations 2. Climate change over the next few decades 3. Stratospheric ozone depletion and increased UV radiation 4. Changes in land cover and in terrestrial and marine ecosystems So-called cross-cutting aspects are also taken into account: - Global observing systems (e.g. Earth Observing System, EOS) - Research on the human dimensions of global change and development of tools for conducting integrated environmental assessments - Education and communication in the area of global change

Table 1 (contd.)

	Great Britain	Japan	Canada
Evaluation/ orientation	Research is conducted primarily in the natural sciences (e.g. oceanic circulation; biogeochemical cycles; ecosystems). Involvement in programs such as GEWEX (Global Energy and Water Cycle Experiment) and GCOS (Global Climate Observing System) plays a major role. Socioeconomic research is mainly carried out by the ESRC (Economic and Social Research Council). The latter operates its own Global Environmental Change Programme (GECP), comprising 79 projects, 21 fellowships as well as 14 one-year Starter Grants with a ten-year budget of approximately US$ 35.8 million. In addition, ESRC operates the Centre for Social and Economic Research on the Global Environment (CSERGE) and the Centre for Study of Global Environmental Change (CSEC).	The program is predominantly oriented to natural sciences and technology. Social science research, comprising three of 39 core projects, accounts for approx. 8% of the total and focuses on the following issues: population growth and developing countries; urban design in a sustainable society; land use for global environmental conservation. With regard to the natural sciences, atmospheric research dominates with 13 projects and main emphasis on global warming and depletion of the ozone layer.	The CGCP is to act as an umbrella program, networked to include both social and natural science research. The aim of the program is to establish the same organizational and support bases for the social sciences and humanities as are applied in the natural sciences. The CGCP is unique in that it has to be viewed as an advisory capacity which coordinates and sustains global change research. The program provides funds and supports research planning in order to improve the infrastructure for research in this area. The CGCP does not coordinate and finance all global change research in Canada. This is done by the Department Environment Canada, among others.
Contact	UK Global Environmental Research (GER) Office David Philipps Building, Polaris House, North Star Avenue, Swindon SN2 1EU, UK ph: 0044-1793-411-779 fax: 0044-1793-444-513 Email: ukgeroff@wpo.nerc.ac.uk WWW-Homepage: http://www.niss.ac.uk/education/rc/ukgeroff.html	Research and Information Office, Global Environment Department Environment Agency 1-2-2, Kasumigaseki, Chiyoda-ku, Tokyo 100 Japan ph: 0081-3-3581-3422 fax: 0081-3-3504-1634	Canadian Global Change Program (CGCP) Royal Society of Canada 225 Metcalfe, Suite 308 Ottawa, Ontario K2P 1P9 Canada ph: 001-613-991-5640 fax: 001-613-991-6996 Email: dhenders@rsc.ca WWW-Homepage: http://datalib.library.ualberta.a/cgcp/

Table 1 (contd.)

	Netherlands	Switzerland	USA
Evaluation/ orientation	The central focus of the program is on climate change. The issues are problem- and policy-oriented. Although the program aims at research into the human-induced causes and effects of climate changes, the social sciences play only a minor role. Of the total budget for the 2nd phase, a maximum of US$ 10 million is available for research fields 3 and 4. In phase 1, social science issues were treated in only 25 of the over 140 projects.	The SPPE is organized throughout along interdisciplinary lines. Single projects are not supported, research projects are combined instead into "Integrated Projects" or "Project Groups". Natural and social science research have equal priority. All research work should contribute to introduction and support of sustainable development in industry and society. The SPPE sees itself as a platform for the targeted international integration of projects, especially within the EU.	Budgeting for natural science research projects, particularly those of NASA (e.g. EOS (Earth Observing Systems) Flight Development with US$ 442.6 million, and EOS-DIS (Data and Information System) with US$ 289.6 million), far exceeds the sums appropriated for human dimensions research. Natural science research primarily encompasses climate and UV research, ecosystem research, the study of biogeochemical processes as well as modeling and monitoring. Social science research is predominantly funded by the NSF (National Science Foundation) with programs on the Human Dimensions of Global Change US$ 19.2 million) as well as Institutes/Education US$ 3.1 million).
Contact	NRP Programme Bureau RIVM P.O. Box 1 NL-3720 BA Bilthoven The Netherlands ph: 0031-30-743211 fax: 0031-30-251932 Email: nopsecr@rivm.nl WWW-Homepage: http://deimos.rivm.nl/	Swiss Priority Programme Environmental Technology and Environmental Research (SPPE) Länggaßstraße 23 CH-3012 Bern Switzerland ph: 0041-31-302-5577 fax: 0041-31-302-5520 Email: sppe@snf.ch WWW-Homepage: http://www.snf.ch	Coordination Office of the U.S. Global Change Research Program Suite 840 300 D Street, S.W. Washington, DC 20024 USA ph: 001-202-651-8250 fax: 001-202-554-6715 Email: office@usgcrp.gov WWW-Homepage: http://www.usgcrp.gov

3   German Research on Global Change (Current Status, Evaluation, Scientific Issues)

3.1
Climate and Atmosphere Research

As the WBGU has never ceased to emphasize in all its Reports, the atmosphere, hydrosphere, lithosphere, pedosphere and biosphere form a coupled system (for an overview, WBGU, 1993). In the section that follows, global change research on each of these environmental compartments is analyzed in separate sub-sections for the sake of simplicity. *Boxes 6 and 7* contain the necessary cross-cutting overview of the various sub-systems.

3.1.1
Relevance of Climate and Atmosphere to Global Change

Climate has obvious importance for humankind and the ecosphere. The spread of human settlements, or the formation of soils and vegetation zones, are largely shaped by climatic factors. In recent decades, there has been a discernible human contribution to global warming (IPCC, 1996), setting in motion a potentially dangerous trajectory of climate change. The expected global warming of around 2 °C over the next 100 years (mid-range climate scenario of IPCC, 1996) would bring about a shift in the zonal patterns of habitats and hence in the patterns of agricultural production. The associated impacts on food security for human populations are still uncertain. The expected average sea level rise of 50 cm as a result of the thermal expansion of the oceans and through the melting of glaciers and ice sheets threatens the inhabitants of island states and coastal regions; this is all the more serious given the high population density in precisely these areas. According to the new IPCC Assessment, weather extremes such as severe storms, floods and droughts will increase in frequency and intensity worldwide. This is one reason why the 1990s have been declared an International Decade for Natural Disaster Reduction (IDNDR).

The composition of the atmosphere is a prime determinant of living conditions on Planet Earth. Factors which play a major role include the stratospheric ozone layer and the oxidation capacity of the atmosphere, which is closely linked to tropospheric ozone. The stratospheric ozone layer protects life on Earth against harmful ultraviolet radiation. Greatly enhanced intensities of UV-B radiation have been measured during the austral spring and early summer (October to December) as a result of the (seasonally varying) depletion of this layer in the Antarctic and neighboring zones of the southern hemisphere. The implications for the biosphere are still largely unknown. In the northern hemisphere, too, "miniholes" in the ozone layer may form under certain meteorological conditions over specific regions or periods of time, in addition to the general background decline.

The atmosphere's oxidation capacity enables it to purify itself of biogenic and anthropogenic trace constituents, accumulation of which would be damaging for humans and ecosystems. Through the use of fossil fuels and the combustion of biomass, trace gases are emitted which undergo a chain of chemical reactions, leading to an increase in the concentration of ozone in the troposphere. Many crops and tree species exhibit a low tolerance to enhanced ozone concentrations, so reduced crop yields can be expected. Moreover, this increase in tropospheric ozone concentrations poses a threat to human health. The long-distance transport of other substances with acidifying and eutrophying effects endangers sensitive ecosystems, especially in the northern hemisphere.

3.1.2
Climate Research

3.1.2.1
Major Contributions by German Climate Research

PALEOCLIMATE

Paleoclimate research findings are important for climate research, especially where they provide evidence for anthropogenic perturbations. Rapid fluctuations in climate have been identified for all periods prior to the present interglacial (Holocene) on the basis of ice cores from the inland ice sheets of Greenland. However, strong fluctuations in the interglacial period approximately 125,000 years ago cannot be interpreted without uncertainties remaining, due to the lack of congruence with neighboring

ice cores. German scientists have made significant contributions in this field.

GREENHOUSE GASES

Changes in greenhouse gas concentrations are monitored by a global network of observation stations (Global Atmosphere Watch, operated by the WMO) in which German research groups are also involved, particularly in the quality assurance field.

GLOBAL MODELS, CLIMATE VARIABILITY AND CLIMATE PREDICTION

Major progress has been achieved in recent years in the study of climate variability. The interpretation of existing data has been improved, while a number of research programs are now conducting high resolution observations of the ocean as well. Concrete proof of an anthropogenic "climate signal" has only been obtained because models tested against these observations are now able to simulate natural variability with reasonable precision (IPCC, 1996). How-

---

BOX 6

Cross-Cutting Issues – Climate Impact Research

RELEVANCE, CHARACTER AND
INTERNATIONAL STATUS

"Some of the key tasks of *climate impact research* are the identification and evaluation of the possible effects of global change on natural systems and civilizations as well as the potential protective and mitigation measures with respect to these effects. ... "Avoidance of causes" as a strategic option is deliberately countered with the contrary position advocating "toleration of and/or adaptation to these effects" (WBGU, 1995b).

This brief description by the Council of the main elements of climate impact research is at the same time a definition of the special interpretative outputs expected from this relatively new scientific field. Its task is to combine and integrate knowledge from the most diverse sectors in order to support all phases of the problem-solving process – from the assessment of relevance to the compilation of recommended actions. Climate impact research is therefore oriented, *per constructionem*, to *horizontal* and *vertical* integration, as dealt with in detail in *Section C*. While there is a certain overlap between research of this kind and climate research in general, its specific methodol-

ogies set it apart as one of the interdisciplinary projects within a "new" science of the environment.

The high expectations placed in climate impact research within the framework of UNCED are illustrated by Article 2 of the Framework Convention on Climate Change, which states that "The ultimate objective of this Convention [...] is to achieve [...] stabilization of greenhouse gas concentrations in the atmosphere at a level that would prevent dangerous anthropogenic interference with the climate system. *Such a level should be achieved within a time frame sufficient to allow ecosystems to adapt naturally to climate change, to ensure that food production is not threatened and to enable economic development to proceed in a sustainable manner.*" (BMU, 1992)

The urgency, scope and design of all international climate protection strategies are therefore dependent on the answers that research provides to the question of tolerable ecological, economic and social impacts. The need for definitive statements about the consequences of large-scale climate changes for the various sectors and segments of global society *at regional scale* is underscored by the prevailing lack of consensus in the impact assessment field, which may be partly due to conflicting political interests.

Major gaps in knowledge exist with respect to the following areas:

▶

- Regional manifestations of global climate change and implications for water supply, water demand and water management, as well as for economic sectors dependent on water (agriculture, tourism, etc.).
- Effects of global climate change on the stability of dominant atmosphere-ocean circulation patterns (conveyor belt, ENSO, Asian monsoons, etc.), and possible impacts on natural and social systems.
- The extent to which the ecological performance of biogeochemical cycles and the structure of ecosystems can be influenced by climate-related perturbations.
- Effects of potential climate changes on human health, particularly through shifting of the geographical zones in which pathogens occur.
- Interactions between anthropogenic climate change and soil degradation caused by land use.
- Climate-related changes in the frequency and characteristics of extreme weather events (storms, floods, droughts, etc.) and the implications these have for disaster prevention.
- Political and sociocultural impacts of climate change at high regional resolutions, with special reference to new or enhanced conflict potential.

These issues are interlinked in a complex network of interrelations, and are exceedingly difficult to respond to if isolated from each other in any way. There is a growing consensus among experts that the development of Regional Integrated Models (RIM) holds out great promise as a climate impact research approach (IPCC, 1996; WBGU, 1995b).

GERMAN ACTIVITIES AND CONTRIBUTIONS
TO INTERNATIONAL PROGRAMS

Germany has realized the importance of these issues and has responded by establishing the Potsdam Institute for Climate Impact Research (PIK). The work carried out at PIK includes sectoral impact analyses, integrated regional studies, Earth System research and methodological studies.

Of the various activities currently engaged in by the German climate impact research community, special reference should be made to two major network projects in particular. One is the joint government-*Länder* program entitled Climate Change and Coasts (Ebenhöh et al., 1995), the other is ICLIPS, a new project with the goal of integrated analysis of global and national climate

protection strategies. The latter project is based on an international research network and the "*crash barrier*" research approach developed by the WBGU (1995b and 1996). The results of these two projects will be integrated directly into international debate through the IPCC.

Other centers conducting climate impact research include the Wuppertal Institute for Climate, Environment and Energy, the Humanity, Environment and Technology (MUT) Program Group at the Jülich Research Center, the Institute of Hydrophysics at the Geesthacht Research Center (GKSS), the Fraunhofer Institute for Systems and Innovation Research (ISI), the Center for Agricultural Landscape and Land-Use Research (ZALF), the Federal Agricultural Research Institute (FAL), and the Institute for Physical and Theoretical Chemistry at the University of Frankfurt am Main.

STRUCTURAL AND INFORMATIONAL
IMPROVEMENTS NEEDED

In light of the general gaps in knowledge referred to above and Germany's national interests and responsibilities, climate impact research in Germany should begin or continue to tackle the following priority issues:

- Climate sensitivity of the agriculture and forestry sectors in central Europe, with special reference to global economic and demographic trends.
- Vulnerability of German regions and economic sectors to extreme events induced by altered climate (e.g. floods).
- Potential threats to our society caused by the knock-on effects (through migrational flows, market turbulence, adaptation of natural systems, etc.) of climate-induced conflicts and damage in other regions of the world.
- Limits imposed by fundamental ethical and esthetic principles on the tolerance of climate change and its impacts.

These complex issues can best be approached in methodological and structural terms through integrated regional models (especially for coastal, mountainous and semi-arid regions) or through specific network projects and priority programs of the BMBF, the German Research Foundation (DFG) and other research organizations. By its very nature, climate impact research cannot be conducted as a mere collection of individual studies addressing specific details of the total system.

ever, there are very few research groups worldwide that construct and operate coupled ocean-atmosphere models capable of projecting future climate through the integration of human-induced climate forcing. Two of these models, one of them at the Max Planck Institute for Meteorology in Hamburg, have succeeded for the first time in modeling atmospheric temperature change to within the range of natural variability observed from pre-industrial times to the present.

The potential benefits to be derived from accurate predictions of climate fluctuations have been clearly demonstrated for the ENSO phenomenon (El Niño-Southern Oscillation, irregular ocean warming in the eastern Pacific) – the TOGA program operated by WCRP has succeeded in predicting occurrences of this phenomenon in South America, eastern Australia and the Pacific island states, thus enabling the socioeconomic impacts, especially in agriculture, to be cushioned through the implementation of preventive measures.

These research efforts have received substantial support from the climate research program of the BMBF. Germany has acquired a leading position in the field of climate modeling thanks to support and funding from the BMBF and the EU. The BMBF is still deliberating on how best to continue its program in this field, but one of the outcomes should be to build on the progress already achieved and to extend the focus to the modeling of global biogeochemical cycles.

REGIONAL CLIMATE MODELS

Local and regional climate is determined through the interaction of large-scale circulation patterns, circulation systems induced by geographical and topographic factors, and smaller-scale processes at the Earth's surface which are subject to human influences. These various processes operate at different scales and must be incorporated into climate models in greater detail than is presently the case. Several institutes in Germany have made outstanding contributions in this respect.

3.1.2.2
Integration of German Climate Research in International Programs

Many of the critical fields mentioned above are the subject of WCRP and IGBP research programs (*see Section B 1*). The German climate research community is well involved in the GEWEX, WOCE, ACSYS, SPARC and CLIVAR projects of WCRP and in IGBP's PAGES, BAHC and JGOFS Core Projects and GAIM integrated modeling project.

German scientists play an increasingly important role in the design and implementation of the various programs. The same applies to the preparation of the IPCC Assessment Reports (IPCC, 1990, 1992, 1996). German researchers participated in the European Ice Core Programs in Greenland, and will continue this work in the Antarctic; they were also involved in the publication of a paleoclimatic and paleoecological atlas of the continents. The German Climate Computing Center (DKRZ) in Hamburg collaborates intensively in comparative international studies of climate models, and has been commissioned with coordinating the work of the leading climate computing centers in Europe.

Data from national weather stations are archived in a climate database by the German Weather Service (DWD), which also produces and disseminates data sets on global precipitation (the Global Precipitation Climatology Centre operated on behalf of the WMO). Remote sensing data, principally from the European region, are processed by the German Remote Sensing Center at the German Aerospace Research Establishment (DLR) in Oberpfaffenhofen. A climate database is being developed at the DKRZ (for model outputs and measurement data, such as the Paleoclimate Database Unit at the University of Hohenheim). A WMO global runoff data center has been established at the Federal Institute of Hydrology in Koblenz.

3.1.2.3
German Climate Research – Research Needs Concerning Global Change

SCIENTIFIC ISSUES

Paleoclimate research has disposed of an astronomically calibrated climate time scale covering the last 5 million years. The available observation material is plentiful on the whole, but still exhibits some geographical gaps. In particular, there is a lack of data from the tropics, the southern hemisphere and some ocean regions. Analytical processing of regional time series is still in its infancy.

Besides the wealth of long-term time series of meteorological data sets, *climate research* can now access relatively long series of data from satellite remote sensing (since 1972). Systematic use of such data should now be prioritized.

The investigation of greenhouse gases – their sources, absorption of radiation, chemical reactions and transformations, and sinks – is of paramount importance. Despite substantial research efforts, there is still considerable uncertainty about the source patterns of anthropogenic greenhouse gases (especially nitrous oxide; IPCC, 1992 and 1996). The direct and indirect feedbacks of climate changes on greenhouse gas emissions, particularly methane and nitrous oxide (e.g. through changes in the water balance of ecosystems and land-use changes) could be highly significant but are not yet capable of assessment. Interactions between the atmosphere and biosphere play a key role in this respect. There is therefore an urgent need for improved applicability of results from individual studies (*see also Box 7*).

Little research has been done on the direct and indirect effects of aerosols and clouds on climate. This calls for systematic field and laboratory studies and the construction of suitable models. Aerosols have not been taken into account by climate models to any degree of precision. The special expertise of German atmosphere research in the field of aerosols, hydrometeors and trace gas cycles should therefore be utilized to maintain a leading international role. The BMBF priority program currently being planned on the climate-forcing effects of aerosols should therefore commence as quickly as possible.

Other processes that must be described in greater detail in climate models relate to the hydrological cycle – for example the formation and optical properties of clouds, the formation of sea ice and oceanic deep-water. Insolation fluctuations, especially in the ultraviolet range, must also be dealt with in greater detail by climate models. The need for research is similarly great with respect to the identification and prediction of weather extremes and regional climate changes.

STRUCTURAL IMPROVEMENTS NEEDED
Satellite remote sensing is particularly important on account of its global coverage. There are already some long time series, such as the data from the NOAA weather satellites since 1978 and the Landsat series since 1972. The systematic use of these data should be continued and brought into the relevant international projects such as GEWEX. In the field of remote sensing, there is a need to improve the relationship between the development of equipment,

---

BOX 7

The Coupled Atmosphere-Hydrosphere-Cryosphere-Biosphere System

The atmosphere, hydrosphere, cryosphere and biosphere are closely linked through biogeochemical fluxes enhanced by anthropogenic activities – in particular by industrialization. Our understanding of the interactions between these sub-systems is still somewhat patchy, however. The processes operating within the system as a whole transcend the individual "spheres", an aspect that was taken into consideration during the establishment of the two interdisciplinary programs WCRP (primary focus on physical processes) and IGBP (biological and chemical processes).

The purpose of climate models is to describe the main energy and material fluxes between the various spheres, and the internal dynamics of those systems. The most advanced climate models (General Circulation Models, GCM) have succeeded in coupling atmosphere and hydrosphere, but are still inadequate as far as integration of the cryosphere and biosphere are concerned. In particular, there is a need for more intensive investigation of the material interactions between terrestrial and marine biocoenoses and their abiotic environment.

SCIENTIFIC ISSUES
There is still a great need for research into the atmosphere-biosphere interface. A good example for such work is provided by the excellent results and standards attained by German research into *forest decline*. Intensive research efforts have focused here on the damage caused to central European forests by anthropogenic airborne pollutants. Research into forest decline has produced some important (although not yet finalized) results thanks to the interdisciplinary research approach that is applied. The expertise thus gained and the methodological progress achieved in field and laboratory investigations could be applied in other regions besides the temperate zones in order to determine the damage caused to natural ecosystems by airborne pollutants. In the tropics and subtropics, long-term studies should be conducted to identify the natural variability of ecosystems and to ensure better understanding of atmosphere-biosphere interactions. In view of the special significance of the tropics and subtropics for global climate processes and the incisive socioeconomic transformations occurring there (especially through economic growth and land-use changes), it is essential that research efforts be stepped up in those regions. In the modeling field, there is a need for further development of process-oriented simulation models (Soil-Vegetation-Atmosphere-Transfer Models, SVAT).

on the one hand, and the need for data processing and faster interpretation, on the other – the space agencies have lent far too little support to the subsequent evaluation of data.

3.1.3
Stratosphere Research

3.1.3.1
Major Contributions by German Stratosphere Researchers

Every year since the end of the 1970s has witnessed a serious thinning of the ozone layer over the Antarctic during the months of September and October. This "ozone hole" has increased considerably in size since it was first discovered. In recent years, the total loss of ozone has amounted to more than 50% of the total column, and up to 90% at higher altitudes around 18 km. As a consequence, much higher intensities of UV-B radiation are measured during the Antarctic spring. Due to the outflow of air masses depleted in ozone from the Antarctic ozone hole into the stratosphere at mid-latitudes of the southern hemisphere, enhanced ground-level UV-B radiation has been measured in those regions.

The total ozone content is declining worldwide (with the exception of the tropics) by several percent per decade. In mid- and higher latitudes of the northern hemisphere, the decline during the winter months may be as much as 8% per decade. Europe is affected to a particularly severe degree due to meteorological factors, because the low-ozone Arctic polar vortex tends to shift in winter and spring towards northern Europe. Very low total ozone values have been measured in recent years; latest research findings suggest that this trend is related to an increase in stratospheric aerosols following the eruption of Pinatubo. The depletion of the stratospheric ozone layer is one of the most serious and potentially dangerous changes to affect the atmosphere.

The German research community has made key contributions to our understanding of these processes. German research groups are conducting investigations into the dynamics of the stratosphere and the impacts of air traffic on the ozone layer using research aircraft, instrumented balloons and ground-based experiments. Paul Crutzen was awarded the Nobel Prize for Chemistry for modeling the chemical processes in the stratosphere.

3.1.3.2
Involvement in International Programs on the Part of German Stratosphere Research

German ozone researchers are involved in virtually all the major international research programs, e.g. SPARC, and in many cases play a prominent role in the planning and execution of the programs. In addition, German scientists are intensively involved in European research programs (e.g. the Fourth Framework Programme of the EU) and in the Network for the Detection of Stratospheric Change Programme (NDSC). Participation in the European programs can only lead to success, however, if there is a solid basis in Germany.

Inland measuring stations form part of the WMO's global observation networks on the composition of the atmosphere set up within the framework of Global Atmosphere Watch (GAW) and the Global Ozone Observing System (GO$_3$OS).

3.1.3.3
German Stratosphere Research – Research Needs Concerning Global Change

SCIENTIFIC ISSUES
Following on from the work of previous years, the second national Ozone Research Program (OFP 2) was launched in 1996. German ozone research is well endowed with personnel resources and research equipment thanks to BMBF support. OFP 2 covers the following main areas:
- Process studies of the chemistry and dynamics of the stratosphere.
- Identification, monitoring and analysis of the processes leading to variability of stratospheric ozone concentrations.
- Development of predictive capacity (formation of the ozone layer, model engineering).
- Solar UV-B radiation.

STRUCTURAL IMPROVEMENTS NEEDED
Ozone research activities in Germany are broad-ranging in scope, distributed as they are among several research institutes, so most aspects of global change relating to stratospheric ozone can now be focused on. Over the next 10-15 years, however, the following research institutes should be retained or established, also in the framework of European networks:
- Laboratory facilities which allow for realistic experiments to verify current perceptions of stratospheric ozone depletion (aerosol research).
- Research and monitoring capacities in Germany and at selected stations in northern Europe for

measuring the total ozone content and its vertical distribution in the stratosphere and troposphere, as well as other relevant parameters, e.g. temperature distribution and solar UV-B radiation.

- Establishment of comparable research capacities at selected sites in Africa, Asia and Latin America.

The computing capacities required for ozone research are of the same dimensions as for climate research, whereby a three-dimensional global circulation model encompassing the dynamic and chemical processes in the stratosphere demands higher computing capacity than a typical atmospheric circulation model.

The coming years will most likely see some important ozone research contributions from *satellite experiments*, which will either be designed in a national framework or, if built by the ESA, will be managed by principal investigators from Germany. These include the European GOME experiment on global ozone distribution and measurement of active halogen compounds on ERS-2, the SCIAMACHY and MIPAS trace gas experiments on the polar platform ENVISAT, and the CRISTA space shuttle experiment for determining mesoscale dynamic processes in the stratosphere.

Various research aircraft are available for ozone research. The commissioning of the high-altitude STRATO-2C aircraft provides the German research effort with a measuring platform, the only one of its kind worldwide, that could lead to major advances being achieved by global research on stratospheric ozone.

3.1.4
Troposphere Research

3.1.4.1
Major Contributions by German Troposphere Researchers

German *troposphere research* is conducted at a high level in the fields of physical and chemical process studies and coupled modeling of transport processes and chemistry at different spatial and temporal scales. German scientists have made outstanding contributions to the identification of possible anthropogenic perturbations in the oxidation capacity of the atmosphere.

3.1.4.2
Integration of German Troposphere Research in International Programs

Research on tropospheric chemistry and atmospheric biogeochemistry are embedded in the IGAC and GCTE core projects of IGBP, which have started work in recent years. These and other topics are also dealt with in various European programs (e.g. the EUROTRAC project in place since 1988 and its successor project EUROTRAC-II, part of the EUREKA program). German troposphere research is well integrated in these projects, with German scientists playing a significant, sometimes leading role in their conception and execution.

Inland measuring stations form part of the WMO's global observation networks on the composition of the atmosphere (Global Atmosphere Watch, GAW). There is also some collaboration with foreign measuring stations (e.g. Cape Point/South Africa, Izaña/Tenerife).

3.1.4.3
German Troposphere Research – Research Needs Concerning Global Change

In order to ascertain an anthropogenic perturbation of the atmosphere's oxidation capacity, it is necessary to monitor characterizing substances (ozone, radicals, etc.) at different latitudes both inside and outside the planetary boundary layer. This issue is closely linked to anthropogenic climate forcing

To assess the extent of human-induced impacts on climate, it is necessary to conduct model experiments in greater numbers and detail. The global distributions of reactive greenhouse gases (e.g. methane, tropospheric ozone) and precursors of ozone (nitrogen oxides, hydrocarbons) are not adequately known, and the evaluation of budgets is similarly inadequate. There is a need for research into the processes operating at the atmosphere-biosphere interface. The German Troposphere Research Program (TFS) commencing in 1996 addresses this aspect, but only for the European context. Other regions should be included in the analysis in future.

Modeling of chemical processes in the atmosphere should be continued with the goal of developing an instrument for assessing the direct and indirect climate-forcing impacts of trace gas and aerosol emissions that can then be used by political decision makers.

Additional measuring activities, especially in the southern hemisphere and from research aircraft, are necessary in order to determine the oxidation capacity of the atmosphere. The data basis can be improved by including short-lived trace constituents in the GAW measuring program, and by extending satellite-based soundings of the atmosphere down to tropospheric levels.

3.2
Hydrosphere Research

3.2.1
Relevance of the Hydrosphere to Global Change

Water resources are of immense importance for life on Earth. The evolution of life was only possible because prevailing temperatures allow the existence of large masses of liquid water. The Earth's climate is determined to a significant degree, within the coupled atmosphere-cryosphere-ocean system, by the global ocean. Atmospheric factors have the greatest impacts on short-term processes in the oceans, whereas the latter exert the decisive control function as far as long-term processes are concerned. Of relevance here is not only the global transport of heat from lower to high latitudes and the deepwater formation in polar regions, but also the importance of the oceans as a sink and source of greenhouse gases and substances important for terrestrial communities. Not only is the global ocean one of the major compartments of the planetary climate system – its ecosystems are also affected by global change, especially in coastal regions.

Freshwater is a fundamental resource for all forms and areas of life and human society. It is a cultural asset, a basic food for all organisms, but also the basis for food production, many industrial processes and for the generation of energy. Water, as a resource and cultural asset, is endangered by many natural and anthropogenic factors. Both wastage and contamination of water lead to a reduction in useable water reserves and to the degradation or destruction of ecosystems (WBGU, 1993). Mounting water scarcity at local and regional level has turned water issues into a problem of enormous magnitude. In the estimate of the United Nations, about 50 countries are now suffering from severe water scarcity.

Water supply and demand are distributed very unevenly worldwide, i.e. some regions have a water surplus (e.g. sparsely populated humid areas) while others are faced with water scarcity (e.g. large cities in arid zones). Technical problems hinder the redistribution of water across large distances. If water problems are to be solved, then it is necessary to resolve not only the hydrological and natural scientific issues at local, regional and global level, but also the economic, social and political problems that are in-

volved. Holistic, integrated analyses and methods for solving the problems are needed, and must therefore be developed by the scientific community.

3.2.2
Marine and Polar Research

3.2.2.1
Major Contributions by German Marine and Polar Researchers

Physical oceanography and marine meteorology are dominated in Germany by two overriding topics: research into the climate functions of the oceans, and the dynamics of the European marginal seas. Work in these fields is aimed at predicting short-term fluctuations in climate and assessing the stress-bearing limits of coastal marine ecosystems. Measurements in the sea, remote sensing and modeling are the three main, complementary methods for such research.

Of special importance for global change research are the large-scale multidisciplinary studies on the importance of the ocean and its biota as sources and sinks of greenhouses gases (CO_2, methane). Marine biological researchers study matter fluxes, on the one hand, and changes in marine communities and biodiversity as a result of climate change and human activities, on the other. Within the marine earth science agenda, paleoclimatic and paleooceanographic issues are of primary importance for understanding long-term changes in marine circulation, as are the geochemical aspects of matter cycles. Core samples from marine sediments and coral reefs form a kind of natural archive from which information can be gleaned about past climatic changes.

As far as issues pertaining to global environmental change are concerned, German marine research has concentrated first and foremost on the North Sea, the Baltic Sea and the Northeast Atlantic. Regular expeditions to the Southern Ocean have also been carried out over the last 20 years. To this is added the more recent studies on climate-forcing and ecological issues in the Arctic Ocean and the South Atlantic, and the major expeditions conducted in the Indian Ocean and the coastal waters of South America every few years. The study of tropical coastal regions is also playing an ever-greater role.

Until now, German polar research has focused primarily on three areas relating to the global climate system. The first of these foci concerns *energy exchange in the coupled ocean - sea ice - atmosphere system*. Because of its high albedo, sea ice slows the absorption of radiative energy by the ocean and the loss of thermal energy to the atmosphere. With the help of state-of-the-art remote sensing of ice, and model-building, these exchange processes can be determined for changed boundary conditions.

The second main focus involves research into the *mass balance of the polar ice sheets and changes induced by climate fluctuations*. The mass balances of the large ice sheets are influenced by both temperature and precipitation. Climate models show that a temperature increase in the atmosphere leads to higher precipitation in the polar regions. Modeling the mass changes of the large ice sheets permits conclusions to be drawn about sea-level changes.

A third focal point of research concerns the *reconstruction of climate using polar archives*. With the help of ice cores, such as those obtained from the European GRIP program in Greenland, it is possible to perform historical analyses of atmospheric and chemical composition, as well as depositions of particulates from the atmosphere. The Greenland cores enable researchers to reconstruct the region's climate history over the last 200,000 years at high temporal resolution. Sediment cores taken from polar oceans and lakes, and from permafrost soils are an additional source of global climate data.

3.2.2.2
Integration of German Marine and Polar Research in International Programs

Marine and polar research is well organized within the international core programs, as well as in numerous individual projects, and the German research community is intensely involved in these activities. Four core projects of the IGBP relate to the hydrosphere in a special way:

- BAHC (Biospheric Aspects of the Hydrological Cycle)
  German research groups are involved in all four priority research areas of this program. The strength of Germany's commitment is evidenced by the fact that the International Core Project Office is located in Potsdam. Closely connected to this program is the Hydrological Cycle priority program of the BMBF, which encompasses a total of 33 projects on The Hydrological Cycle in Climate Models.
- GLOBEC (Global Ocean Ecosystem Dynamics)
  The German contribution to this new core program (established in 1995) is currently being formulated. The project studies the regional impacts of global climate and circulation changes in regional studies, as exemplified by simultaneous changes in fish stocks in different parts of the world ocean (Large Marine Ecosystem concept, LME, e.g. Baltic Sea and Benguela Current).

- JGOFS (Joint Global Ocean Flux Study)
  Since 1990, the BMBF and the DFG have been financing Germany's activities in this project, both at national and international level. Various Special collaborative units are integrated within JGOFS, for example SFB 313 Environmental Changes in the North Atlantic at the University of Kiel and SFB 261 The South Atlantic in the Late Quaternary: Reconstruction of Mass Balance and Current Systems at the University of Bremen.
- LOICZ (Land-Ocean Interactions in the Coastal Zone)
  German contributions to LOICZ are concentrated on the North Sea within the KUSTOS project (Coastal Material and Energy Fluxes), the Baltic Sea in the research network for the Mecklenburg-S.W. Pomeranian coastal landscape (GOAP, TRUMP, ÖKOBOD) and on the dynamics and management of mangroves (MADAM) in a joint German-Brazilian project. The multilateral Red Sea Program supported by the BMBF has also connections to LOICZ, as well as the program Climate Change and Coastal Zones (*Box 8*).

WCRP accounts for two other core programs in this field:

- GEWEX (Global Energy and Water Cycle Experiment)
  BALTEX (Baltic Sea Experiment), a sub-project of GEWEX studying the water balance of the Baltic catchment, receives special support from Germany. The International Secretariat of this program is located in Geesthacht.
- WOCE (World Ocean Circulation Experiment)
  This program is being carried out over the 1990-1997 period and involves a strong German presence in all three Core Projects. The TOGA program (Tropical Ocean-Global Atmosphere), in which German researchers made a substantial contribution in the form of ship-borne measurements, has already been completed.

3.2.2.3
German Marine and Polar Research – Research Needs Concerning Global Change

SCIENTIFIC ISSUES
Although German marine and polar research covers a broad range of topics, partly due to the variety of institutions involved, it has been predominantly geared to natural scientific issues. The focus on other areas, such as land-based marine pollution, anthropogenic perturbations of food chains and species diversity, or the structure and functioning of international environmental regimes (for example the Convention on the Law of the Sea) is still relatively new and un-

BOX 8

Cross-Cutting Research on "Climate Change and Coastal Regions"

The fact that coastal zones are the most densely populated and the most intensively used regions in the world means that climate change operates here on socioeconomic and geographical structures of relatively high vulnerability. In order to assess this risk potential for Germany's coastal areas, the former BMFT initiated a research program in 1991 on Climate Change and Coastal Regions to study the potential damage to and vulnerability of coastal areas, seen as habitats and economic zones characterized by manifold interlinkages and interactions.

The North Sea and Baltic Sea coasts are greatly affected by human activities. Their sensitivity to climatic effects has already been severely altered as a result of other impacts, such as depositions of nutrients and pollutants, groundwater extraction, artificial deepening of estuaries, etc. For this reason, the research program focused not only on susceptibility towards storm floods or other natural disasters, and changes in marine and littoral ecosystems, but also involved an integrated analysis of the potential exacerbation of existing conflicts over water resources (especially between agriculture, coastal protection and tourism).

Climate Change and Coastal Regions, a joint government-*Länder* project, considers a broad spectrum of climate impacts specific to coastal areas, analyzes the stress-bearing capacity and elasticity of many sub-systems, and includes assessments of future trends within natural and social systems. A panel of experts has developed an integrative research concept based on new methodologies: modeling hydrographic, morphological and biological processes; coupling natural and social scientific models; holistic and policy-based analysis of selected areas (case studies on the Weser Estuary and the island of Sylt), as well as the identification, description, and evaluation of the resultant threats and conflicts by means of a Geographical Information System (GIS).

The research priorities of the program are: the analysis of historical changes in climate and coastal morphology; studying the impacts of climate on the structure, dynamics and stability of current systems; sediment transport; changing storm flood patterns, and extreme hydrological and wind events. Building on these, there are also studies on coastal protection, the economic risks associated with broken dikes, floods and high winds. The range of topics also includes the potential risks and damage at the North Sea and Baltic Sea coasts for a sea-level rise of 1 meter, analyses of public perceptions of extreme weather events, and the response and adaptive behavior of socioeconomic systems.

At international level, this research program forms an integral part of the Coastal Zone Management Programme of the IPCC, which studies the global significance of coastal zones in the context of sea-level rise and other climate impacts. Using the methodologies and scenarios developed by IPCC, standardized risk assessments are carried out worldwide to provide the scientific basis for appropriate action and response strategies in coastal regions to actors in the political and economic domains.

The experience gained through these studies of the North and Baltic Sea coasts should be extended to cover selected coastal and island regions in the tropics and sub-tropics, as demanded by AGENDA 21. An initial basis for this could be the joint German-Brazilian research project in the Amazon delta, MADAM, and the contact unit for tropical coastal research, both of which are supported by the BMBF.

---

developed. These and other issues require close collaboration between different disciplines in the natural and social sciences, and in the view of the Council are essential for our understanding of marine environmental changes and their implications for global change. This means that there are additional research needs in this area.

The main problem areas that should be tackled more intensively within the European Union over the next 10-15 years can be summarized as follows:

- Expansion of the scientific basis for a Global Ocean Observing System (GOOS), analogous to the WMO's Global Weather Observation Network.
- Research into the Arctic Ocean, with special reference to climatological aspects.
- Research into deep-water regions and the deep-sea ocean floor as a habitat sensitive to anthropogenic influence.

- Research into human-induced impacts on marginal seas and coastal zones.
- Development of indicators for anthropogenic changes in the carrying capacity of coastal ecosystems.
- Technology impact research into the exploitation of marine resources (offshore technology, aquaculture, etc.).
- Development of the scientific foundations for integrated management of coastal regions.
- Extension of hydrological models to include socioeconomic and demographic aspects (e.g. linking up with the BALTEX project).
- Development of dispute settlement procedures for international conflicts over the use of marine resources.
- Research work for assessing marine environmental protection programs and the UN Convention on the Law of the Sea.

Some of these topics require new methodological foundations and instruments, which have to be developed through a joint effort on the part of industry and the research community. The Federal Government's 1993 marine research program, the new Polar Research Program, the Deep Sea Research Concept and several BMBF research projects on climate and the Baltic Sea have already approached these issues to a certain extent, but in the estimation of the Council they should also endeavor more to promote and integrate social scientific research on the environment. A good example at international level of how such a concept could be put into practice is provided by the Large Marine Ecosystems (LME) project, which studies large marine areas such as the North Sea or the Baltic, or current systems like the Benguela and Peru Currents with respect to their productivity, food chain structure and exploitation in the light of climate change and anthropogenic stresses.

STRUCTURAL IMPROVEMENTS NEEDED

If those fields of German marine and polar research relating to global change are to be shaped and organized more efficiently, what is needed is not the establishment of important new institutions or committees, but bundling and organizational reorientation. This would include improved use of shipping capacity combined with stepwise modernization of the research fleet, as well the promotion and integration of social scientific environmental research mentioned above.

In the field of measurement and surveillance technology and remote sensing, German institutes and companies must cooperate more closely not only within the European framework, but also at international level. This may require a control mechanism of some kind. The same applies to the promotion of interdisciplinary collaboration, especially between natural and social science disciplines.

In general, the benefits provided by the variety of research establishments in Germany, including those covering areas of marine and polar research with a global change component, should be enhanced further by including funding instruments for joint projects, whereby greater attention must be placed on interdisciplinary networking than has been the case so far. In this connection, the Council recommends that a DFG priority program be set up for the field of coastal research.

3.2.3
Freshwater Research

3.2.3.1
Major Contributions by German Freshwater Research

Freshwater research, and especially hydrological research, is relatively well developed in Germany. The main bodies that are active here are universities, the Federal Institute of Hydrology and various regional authorities for water resources management. However, there is no central, coordinating research establishment for freshwater research with a global perspective.

Besides the established hydrological research activities in state institutes, there also exist a large number of smaller to very small working groups that take a problem- and solutions-oriented approach to water demand and water protection issues. These include the *Öko-Institut* (Eco-Institute) in Freiburg, the *Institut für Sozialökologische Forschung* (Institute for Socio-Ecological Research) in Frankfurt, and the *Institut für Strömungswissenschaften* (Institute for Flow Research) in Herrischried. All of the latter are endeavoring to develop new interdisciplinary approaches and methodologies. Within the framework of the Urban Ecology research program, the BMBF is supporting two network projects focusing on unconventional technological solutions (using rainwater), and social science research on the behavior of actors (*water culture*).

There are tentative beginnings in Germany of a broader research focus centering on the role of water resources as a cultural asset. Work in this "outsider" field has been dominated so far by studies on historical achievements in the field of water management and water engineering, the symbolic importance of water in historical cultures, the cultural importance of river systems (e.g. the River Rhine) or the loss of local and regional "water culture".

3.2.3.2
Integration of German Freshwater Research in International Programs

The German research community has demonstrated a powerful commitment to international research programs addressing freshwater issues. Various National Committees have been formed to coordinate efforts (such as IHP/OHP, WCRP/GEWEX, IGBP, IDNDR). Germany was also chosen as the location for the International Secretariats of the BAHC, PAGES and BALTEX projects (*see Section B 1.2*).

3.2.3.3
German Freshwater Research – Research Needs Concerning Global Change

SCIENTIFIC ISSUES
International efforts to model the hydrological cycle at regional to continental scale have been stepped up within the context of the climate change debate. However, there is still a substantial need for research to refine the natural scientific dimensions of these models and to link them to other models. This is important above all for the calculation of precipitation by climate models and in the analysis of regional water resources.

Special attention must be devoted to the effects on water resources arising from significant changes in natural and societal boundary conditions. Key research issues concerning global change include floods and droughts. "Probabilities of Extreme Hydrological Events" and "Declining Quality of Freshwater Resources" are thematic complexes of crucial relevance for the populations affected. Research aimed at elaborating methods for sustainable water use and capacity building is helping to avoid or mitigate the problems experienced by the regions worst affected.

The Council classifies the basic global water problems in three categories: *scarcity, pollution* and *wastage* (WBGU, 1993). In view of the fundamental importance of water for sustainable development, the Council identified four central fields of action: *water demand, water supply, water pollution* and *natural risks*.

As far as the *demand for water* is concerned, there is no doubting the many unexploited options for handling water resources more carefully, i.e. for reducing water consumption and/or increasing water productivity. This calls for both efficient use of water resources and appropriate water-saving technologies. One prerequisite is the metering of water consumption. Suitable methods must be adjusted to local and regional circumstances. Research must also supply the scientific foundations and recommended solutions for careful handling of resources and for demand side management.

Water stocks can be increased in a variety of ways, including conventional, non-conventional and new, untested methods (*supply side management*). Research is needed on ecologically effective, economically efficient and socially acceptable ways of increasing water stocks.

With regard to *water resource protection*, there is a need to intensify efforts worldwide in order to prevent the pollution of surface water and groundwater stocks and to reduce the resultant health hazards. Research in this area is directed primarily towards the development of processes and procedures for monitoring the quality of water. The quantitative surveying of water resources must play a greater role in future, alongside the traditional field of water analysis.

The fourth area of activity can be described as *disaster management*. The frequency and intensity of severe floods and droughts have increased over time, inducing large-scale regional migration (e.g. Bangladesh, Somalia, Sudan). Research policy must place greater emphasis not only on emergency aid, but also on preventive disaster management, i.e. on improved adaptation to and advance preparation for such events.

GLOBAL WATER STRATEGY
The research components of a *global water strategy* must be integrated into a systematic analysis of the causes and effects of the various water problems that are expected to arise in the future. A whole series of bilateral and multilateral initiatives are conceivable for implementing such a strategy, and a high priority for research must be to integrate such efforts within a consistent approach. To this end, it is necessary to define the relevant goals, develop appropriate instruments and create effective institutions.

Following on from previous recommendations made by the WBGU, the need for global change research pertaining to freshwater resources can be summarized in the following keywords, whereby an implicit aspect is the integration of natural and social science research activities:

- *Inventorying of water resources:* determine water availability; determine intensities of water consumption with respect to sectors, regions and products ("water eco-balances"); improve the measurement and description of biosphere-hydrosphere exchanges (hydrological cycle); develop regional water resource models, coupled to climate, vegetation and anthroposphere, for identifying the impacts of changes in climate and land use, and of measures in the field of water management.

- *Water efficiency and water saving:* develop low-consumption technologies for drinking water supply, irrigation and industrial production as well as methods for water saving, and analyze these with respect to their ecological side-effects; refine models for integrated resource planning in the water sector, including methods for demand-side management.
- *Water culture:* analyze cultural values relating to water; disseminate knowledge gained from experience and practical learning approaches.
- *National water policy:* comparative assessment of optimal water policies: goals, instruments (price- and volume-based solutions) and institutions (private and collective water rights, regional water associations). Research into traditional methods of water management with special reference to sustainability and possible transfer to modern institutional set-ups.
- *International water policy:* analyze experience with transboundary water management systems and communicate solutions to existing conflicts; prepare and support a pilot project on "water partnerships" between Germany and some developing countries; integrate approaches existing in Germany, such as precipitation climatology, global runoff data register, etc. in a national database and/or International Water Institute.

STRUCTURAL IMPROVEMENTS NEEDED

The cardinal objective for the future must be to coordinate research work in the diverse regions and disciplines and to organize cooperation between the respective institutions, disciplines and research levels in order to achieve specific goals. Examples include:

- the DFG priority program entitled Regionalisation in the Field of Hydrology,
- The Hydrological Cycle priority program and the BMBF's network project on Urban Ecology.

However, these objectives can only be achieved if additional resources are provided for hydrological research in Germany.

Compared to other European countries and other research disciplines, the German research community is under-represented in international bodies for hydrological research. This is the case not only for hydrological issues, but also for economic and technological aspects.

In the view of the Council, water resources are of such vital importance to humanity and the threats to these resources so serious from the global perspective that it intends to deal with global water issues in one of the forthcoming Annual Reports. Special attention will be dedicated to hydrological research in the widest sense and to its links with other research fields. The Council will then take a position on structural and organizational issues, e.g. with respect to a national Water Resources Institute in Germany. Such an institute could assume responsibility for national and international coordination, and integrate hydrological aspects of the ecosphere and anthroposphere in the framework of joint research programs.

3.3
Soil Research

3.3.1
Relevance of Soils to Global Change

Soils are a component of terrestrial ecosystems and thus perform an important function as a habitat for organisms (*habitat function*) and as a regulator of matter cycles (*regulatory function*). By performing these functions they ensure the recycling of nutrients and the supply of water to primary producers, i.e. green plants. As structural and functional elements of terrestrial ecosystems, soils exchange energy, matter and information with the atmosphere, hydrosphere, biosphere and lithosphere, and are thus exposed to changing environmental conditions.

Soils are subject to constant change through the weathering of minerals and the input or output of substances along with water and air, but also through the immigration and emigration of organisms. These processes occur very slowly under natural conditions – at centennial to millenial time scales – so that biological communities are able to adapt to these conditions and in many cases even mitigate degradation processes. This results in terrestrial ecosystems that are relatively stable despite natural variations in boundary conditions.

The fact that soils serve as the site for plant growth and supply the latter with nutrients and water makes them an important production factor in agriculture and forestry (*utilization function*). As a result, soils are subjected to constant land-use changes that often induce various types of degradation. In addition to their role as a production factor in the production of food and fodder, soils also perform a *social* and *cultural function*.

Due to rapid growth of the world population and the concomitant intensification of agricultural activities, the environmental conditions for terrestrial ecosystems are now changing so rapidly that the capacity of their endogenous mechanisms for compensation and restoration are exceeded (*see Box 9*). Moreover, use-related interference with soil structures and soil processes causes irreversible changes that can lead to permanent disruptions in the habitat, regulatory, utilization and social/cultural functions of soils.

In its 1994 Annual Report, the Council focused attention on the vulnerability of soils and on the disastrous impacts of soil degradation as one of the main trends of global change (WBGU, 1995a).

3.3.2
Major Contributions by German Soil Researchers

Germany has a long tradition of soil research, but this has mainly centered on how to secure and increase yields in agriculture and forestry (*utilization function*). Only in the last two decades have soil scientists begun to focus on the other soil functions as well. The key issues in German soil research relating to global change are:
- Behavior of natural and anthropogenic substances (carbon, nitrogen, sulfur, organic and inorganic extraneous substances) in soils and their transfer to neighboring terrestrial and aquatic ecosystems and to the atmosphere.
- Impacts of chemical stresses on the structure and function of biological communities (microorganisms, fungi, flora, fauna – *ecotoxicology*).
- Surface sealing and fragmentation of landscapes.
- Use-related loss of soil organic matter and soil compaction, and associated loss of soils through wind and water erosion.
- Assessment of the impacts of climate and land-use changes on soil structure and soil functions (*indicator systems*).
- Importance of soils for species and habitat conservation.
- Application of Geographical Information Systems (GIS) and simulation models to soil properties and processes at the regional level, and to trends in their development.
- Importance of soils for the matter and energy turnovers of ecotopes and landscapes (river, forest, agrarian, urban and natural landscapes).
- Response strategies for the conservation and restoration of soils (soil protection and remediation).

3.3.3
Integration of German Soil Research in International Programs

Traditionally, the main focus of German soil research has been directed at solving problems at national level. This is certainly one reason for the insufficient extent of German involvement in international programs such as GCTE, BAHC and IGAC. However, those researchers who have actively participated in these projects have sometimes played a decisive role in the planning process. Lack of funding has been a major factor behind this low level of German involvement.

A number of individual researchers and research groups are working on soil problems in regions outside Europe. Examples include the various studies funded by the DFG, the BMBF and the GTZ in the fields of agriculture, forestry and the earth sciences.

3.3.4
German Soil Research – Research Needs Concerning Global Change

3.3.4.1
Scientific Issues

In its 1994 Annual Report, the WBGU formulated comprehensive recommendations pertaining to soil research and global change. Accordingly, the following issues in soil research should be assigned priority over the next few years.

REACTION OF TERRESTRIAL ECOSYSTEMS TO CHEMICAL AND PHYSICAL CLIMATE CHANGES
Anthropogenic depositions are causing changes in the composition of terrestrial ecosystems. The complex stuctures of these systems do not yet permit any reliable predictions to be made regarding their structural and functional responses to such loads. In this field of research, further investigations must be conducted into the responses of plants and ecosystems to enhanced levels of O_3 and CO_2 and to increased depositions of acids, nitrogen and other contaminants. The central research issues are:
- Quantification of the long-term impacts of changed global carbon and nitrogen cycles and acidification agents on soils, natural ecosystems and agricultural production (especially the combined impact of enhanced CO_2 concentrations and nitrogen eutrophication).
- Importance of soils as sinks for carbon and nitrogen.
- Degradation of soils and ecosystems due to acid depositions and contamination.

Soil research must cooperate closely with agriculture and forestry, botany and hydrology to elucidate the ways in which terrestrial ecosystems can adapt to global climate changes (water and temperature). In addition to physiological studies on individual plant species, analysis must center on the water exchange of entire ecosystems and regions with the atmosphere and groundwater. Central research issues are:

Agricultural Ecosystem Research on Food Security

Worldwide, 780 million people, or one in seven of the Earth's population, are undernourished. The FAO has forecast that in the year 2010, around 300 million people in sub-Saharan Africa alone will be without an adequate food supply. In theory, there is no shortage of food at present because global food resources suffice to provide each person in the world with the 2,700 kcal that he or she needs to survive. The problem is one of distribution – the food situation in most industrialized countries is characterized by over-abundance, whereas food shortage is rife in many developing countries. Just how rapidly the conditions can change is shown by the shortage of world grain stocks and the resultant doubling of grain prices in 1996, which compelled the EU to impose an export ban on cereal crops.

In the year 2025, about 8 billion people will need food, 2.3 billion more than today (WBGU, 1996). To provide everybody with a food supply equivalent to that enjoyed by the industrialized countries, yields would have to be multiplied by a factor of six. Securing adequate food supplies for the global population is therefore a colossal challenge for future society that can only be mastered if cultivated soils are preserved and/or restored.

Increases in food production were achieved in the past by extending the area of cultivated land and by raising productivity (e.g. by deploying fertilizers, plant protection agents, agricultural machinery and new crop varieties). These methods had negative impacts because newly cultivated lands were increasingly unsuitable (e.g. land producing marginal yields, extreme slopes). The latter trend is visibly linked to population growth in developing countries. Inappropriate methods for mechanical tillage and irrigation, mono-cultivation and excessive use of fertilizers and pesticides to grow High-Yielding Varieties (HYV) have caused mounting long-term environmental damage in many regions. Overexploitation of cropland is accompanied by the mounting threat of nutrient depletion and soil erosion (*see Green Revolution Syndrome, Section C 2.2*). This is compounded by

the destruction of fragile ecosystems in areas where overgrazing occurs.

Ensuring food security, combating poverty and protecting the environment are three intricately related objectives. A central component of rural development efforts must therefore be the promotion of production methods specially adapted to local conditions. In view of the limited options for increasing the amount of cropland, the key to increasing food production lies in improved productivity of existing acreage. However, the risks involved must be kept within the limits dictated by the sustainable development principle. Agricultural research must therefore develop locally appropriate, sustainable and environmentally sound cultivation methods (mixed cropping, crop rotation, intercropping, tillage methods and irrigation techniques) for increasing food production. Modern production technologies (fertilizers, seeds, mechanization) are as important as ever. Applying these techniques and methods is dependent on the requisite social conditions being created as well, however.

Solving these problems is often impossible at local level, so the international community is called on to make greater efforts and provide assistance where needed. There has been a shift of focus in research activities on food security from detailed aspects to the study of food systems in general. Research fields with major relevance for global change are:

- Combining traditional methods of cultivation and pasture farming based on conservational principles with technological innovations.
- Analysis of the adaptive capacity of regional agricultural production systems with special reference to expected climate change and the concomitant shifts in agroecological zones that may ensue.
- Opportunities and risks of biotechnology for food production.
- Integration of production, entitlement and crisis theory approaches in a systems-oriented analysis of global food security.
- Case studies pertaining to the selection of appropriate indicators for determining the vulnerability of social groups and regions to food crises.

- Shifting of vegetation zones as a result of climate changes, with special reference to different soil properties.
- Assessment of the degradation potential of soils as a result of desertification processes (*see Box 10*).
- The role of soils as regulators of the hydrological cycle (estimation of water storage and water infiltration functions) at local, regional and global level.
- Stresses imposed on aquatic ecosystems by eroded soil material.

Changes in Biogeochemical Cycles Caused by Land-Use Practises

The transformation and storage of carbon, nitrogen, phosphorus and sulfur in the soil of terrestrial ecosystems is regulated by biological communities. Land-use related interference with these communities alters soil biogeochemical processes in ways that are difficult to assess. Central research issues are:

- Quantification of land-use related changes in biogeochemical cycles of C, N, P and S, and the processes which regulate these cycles.
- Quantification and regulation of the release or uptake of greenhouse gases from different soils for various types and intensities of use.
- Degradation of soils as a result of the decoupling of matter cycles through land-use changes, and the importance of these processes for the sustainability of land use.

Changes in the Structure and Function of Organism Communities

In particular, there is a need for improved knowledge about the stress-bearing capacity of soils with respect to pollutants and nutrients. Greater research efforts are needed here on the significance of soil organisms and their diversity for the synchronization of metabolism, the decomposition and toxicity of pollutants and hence for the stability of terrestrial ecosystems. Close cooperation with environmental (micro)biologists is essential in this context. Central research issues are:

- Development of methods for classifying and localizing biological communities in soils.
- Ecophysiological assessment of the biotic turnover of matter in soils due to changes in land use and climate.
- Ecotoxicological effects of inorganic and organic pollutants.

Importance of Biodiversity for the Stability and Development of Cultivated Lands

A growing world population will impose much greater demands on soils than is the case even today. If natural ecosystems are not to suffer as a result, then ways must be found to use farm lands in a highly intensive way without causing soil degradation or reducing the stability of ecosystems. Biotic diversity at biotope and species level, which forms the basis for a given landscape's capacity to repair damage and cope with stress, is especially important in this context. Central research issues are:

- Examination of the role played by the heterogeneity of soils and their biological communities for the function and stability of landscapes.
- Strategies for locally appropriate land use based on matter balances.
- Economic and social conditions and consequences for locally appropriate land use.

Earth Observation Systems and Terrestrial Ecosystem Models

Measurement techniques, models and information systems must be developed in this field in order to scale up site-related or patchy information to larger spatial units. Research activities of this kind should be carried out both for the temperate, boreal and for the tropical/sub-tropical zones. Specifically, improvements in comprehensive soil surveys are needed, and the worldwide observation, information and research networks must be expanded.

3.3.4.2
Structural Improvements Needed

Research Organization

In general, German soil science institutes have well-equipped laboratory resources for conducting both quantitative and qualitative analyses of samples (soil, water, plants) to the requisite standard. This is where German soil research could make additional contributions – documenting and disseminating its expertise and experience in soil monitoring and soil analysis. However, the computing facilities currently available are still primitive compared to the modern equipment used in the USA, Great Britain, the Netherlands or Sweden.

The research centers currently in existence are mostly small or even very small working groups. The consequence is that the diversity of specialists required for any systemic approach is often absent. In the past, such diversity was only achieved through research networks. These had to surmount geographical obstacles and the time limitations imposed by

BOX 10

Desertification Research

An extreme form of soil degradation is *desertification*, defined as soil degradation in arid zones resulting from various factors, including human activities. This form of degradation endangers the livelihood system for an increasing number of people, making it one of the central problems of global change. A large part of the scientific community has come to realize that statements on the extent, seriousness and dynamics of worldwide desertification are based on inadequate data. It has been recognized that desertification is not a uniform phenomenon worldwide, and that the causes and effects vary significantly.

Natural geographical aspects have been researched better than socioeconomic factors, but more work needs to be done. In Germany, too, priorities for desertification research have centered on the natural sciences, while research on the various interactions has concentrated above all on the impacts of climate and land-use change on vegetation and landscape development. Although research in these areas is as important as ever, it must be supplemented in future by research in the economic, social and political fields.

Scientific Issues

There is still considerable controversy over the definition of desertification, and to what extent it is increasing in scale. It is crucially important, therefore, that more research be conducted with respect to the identification, forecasting and evaluation of desertification. The same applies to the question of the irreversibility of desertification processes. There is also a need for greater efforts to develop specially adapted programs for combating desertification. Most projects aimed at combating desertification have been (co-)financed with funds from bi- or multilateral development cooperation. *Cost-benefit analyses* must provide the criteria for the prioritization of future research activities and projects. This involves determining the direct costs (income lost as a result of soil degradation) and the indirect costs (for re-

pairing damage to soils). Economically efficient projects are often continued on an independent basis once development projects have run their term.

Another important research issue concerns the optimization of political measures aimed at combating desertification. After a total of eight Preparatory Conferences for the Desertification Convention, there is mounting criticism of the convention, also from environmentalists and development experts. Combating desertification, they say, calls for very different solutions depending on the specific region and sociocultural background; a *global* convention is inappropriate given the complexity of contributory factors. Research on the convention process as it has unfolded to date is likely to produce some essential findings regarding the prerequisites and design of successful environmental conventions.

Structural Improvements Needed

Much is already known about the human activities which cause desertification – namely nonsustainable fuelwood collection, overgrazing, intensive agriculture and inappropriate irrigation methods (WBGU, 1995a) – and there are numerous research projects, also at local level, which address these issues. The main problem here is the way in which research is organized: the anthropogenic factors contributing to desertification exhibit local peculiarities with respect to both natural and sociocultural conditions. To identify interdependencies and outline possible solutions, it is necessary to have in-depth knowledge about the operation of these causes at local level. This requires greater promotion of research at local level, whereby the problem of desertification can only be approached from an interdisciplinary perspective. If efforts to combat desertification are to have any effect, they must be based on a comprehensive approach. The WBGU has formulated a proposal in this respect (*see Sahel Syndrome, Section C 6*). Strengthening desertification research has key relevance to Germany's application to locate the Secretariat of the Desertification Convention in Bonn.

having to acquire external funding. Only recently have there been changes in this respect, with the BMBF playing a major role. Noteworthy examples include the ecosystem research centers and the establishment of soil science institutes at the various

Helmholtz Centers. The MPG and the FhG have realized the deficits in this field and have also set up facilities to address these issues.

Soil research in Germany has been predominantly located in the earth science, agriculture, horticulture and forestry faculties of the universities. Important new institutions focusing on soil research have been created through the establishment of ecosystems research centers (in Kiel, Göttingen and Bayreuth) and through the restructuring of Helmholtz Centers to allow for more intensive ecosystem and environmental research. These have good and sometimes excellent research capacities, being well-endowed with equipment and other resources. The ecosystem research centers have established a presence in the universities. There are Helmholtz Centers at the GBF in Braunschweig, the GKSS in Geesthacht, the UFZ in Leipzig, the KFA in Jülich, the FZK in Karlsruhe and the GSF in Neuherberg. Soil research is also conducted at the Federal Research Institute for Agriculture (FAL) in Braunschweig and at the Center for Agricultural Landscape and Land-Use Research (ZALF) in Müncheberg, as well as in collaborative research centers and graduate colleges in the earth science, agriculture and forestry fields. Mention must also be made of the interfaculty working group Groundwater and Soil Protection in Karlsruhe and the Center for Soil and Water Protection, Regional Planning and Environmental Law in Bonn. Other federal and regional government establishments address soil- and environment-related problems and therefore global change issues. The financial basis for university research is partly provided by the DFG. Integrated research at university research centers and the Helmholtz Centers now receives greater support from the BMBF, and soil studies are also funded by the *Bundesländer* and the EU.

The establishment of the various ecosystem research centers, PIK, ZALF and the SHIFT program, not to mention the focus on ecological issues on the part of the Helmholtz Research Centers, has led to a concentration of research activities in recent years. Links between the various activities of the ecosystem research centers have been created through the TERN research network. The data and analyses produced at these research centers should flow into the IGBP's holistic analysis and in this way advance our understanding of global environmental changes.

The soil research sector in Germany is well-endowed in terms of personnel and equipment. However, it will be faced with grave problems in the near future if financial support is curtailed any further. Soil and ecosystem research work must be conducted in a stable and sustained manner. A crucial advantage enjoyed by university and state research centers is that scientists from different disciplines are able to collaborate, hence ensuring a greater flow of information between the various fields. The German Science Council, for example, has expressly commended the successful cooperation between the GSF and the Technical University of Munich in the Munich Agri-

cultural Ecosystem Research Association (FAM). Cooperation between different research groups also plays a key role in global soil research, but this requires that working groups be able to collaborate on a long-term basis. The translation of current ecosystem research findings into practical action will have to be speeded up in future, so the relevant frameworks – including funding – must be made available.

Remote sensing technologies for measurement and monitoring will have to be deployed more intensively in German soil research, alongside conventional ground-based studies, than has been the case so far. There is also a need to coordinate such activities with international monitoring programs. The research priorities specified above should not only be tackled at national level, but should also be integrated within international projects. The multiplicity of approaches that have developed within the German research system should be maintained as far as possible. In future, however, German soil research must make greater efforts to shape joint international projects, i.e. initiate them and assist in their implementation. The collection and analysis of data must be adjusted to take account of the latest advances, in cooperation with the relevant sectors of the economy. This applies in particular to remote sensing and data processing.

INTERNATIONAL PROGRAMS

German expertise in the field of soil research can help solve the problems faced by other countries; efforts in this direction should place special emphasis on the habitat, regulatory, utilization and social/cultural functions of soils.

Cooperation with Central and East European Countries (CEEC) in the field of soil remediation should be intensified in order to reduce the levels of soil contamination, which in some cases are extremely high. Other regions of the Earth also exhibit extreme physical and chemical loads on soils, remediation of which is beyond the financial and technical capacities of the countries in question. The answer must therefore lie in bilateral or multilateral cooperation. The Common Forum on Contaminated Sites in the European Union set up in December 1994, together with the Concerted Action on Risk Assessment on Contaminated Sites constituted in March 1996 within the framework of the EU's environmental research programs, provide an important basis for priority research activities focusing on the remediation and restoration of contaminated soils.

Combating desertification requires more intensive networking of research with developing countries affected by this problem. Cooperation in the food and agricultural field is of special significance in such regions. In the WBGU 1995 Annual Report, reference was made to the need to strengthen *in situ* research (*see Box 10*).

The Working Group for Tropical and Subtropical Agrarian Research (ATSAF) and the German Association for Technical Cooperation (GTZ) have developed priority issues for desertification research, in close collaboration with the German Foundation for International Development (DSE). A special contribution is being made by the German Alliance for International Agricultural Research (AIDA), which is studying the management of resources in the soils-flora-fauna system.

3.4
Biodiversity Research

3.4.1
Relevance of Biodiversity to Global Change

Biodiversity (or biological diversity) is a general term for the totality of living organisms with all their individual characteristics and interrelationships (Heywood and Watson, 1995). It encompasses the entire spectrum of diversity and variability between systems and organisms at different levels as well as the structural and functional relations within and between these levels, including human action:

- *Ecological diversity* (diversity of biomes, ecosystems and habitats, down to the level of ecological niches).
- *Diversity between organisms* (diversity of taxonomic groups, ranging from phyla, families and genera to species).
- *Genetic diversity* (diversity within and variation between populations, ranging from individuals to genes and nucleotide sequences).

Research in such a complex field must embrace the methods and issues of a wide range of diverse scientific disciplines. These include not only biological sciences (such as molecular biology, conservation biology, agriculture, forestry and fisheries), but also the legal, economic and social science fields (such as law, regional planning, applied social sciences, economics, political science and ethics). Modern biodiversity research should therefore be based on networking between the natural and social sciences and between fundamental and applied research, and should be integrated in the UNCED process (particularly the Convention on Biodiversity and AGENDA 21).

Biodiversity research is relevant for global change because of the values attributed to "biodiversity" as an asset and from the threats to which it is exposed. What is meant here is not only the *intrinsic value* of biodiversity, but also the values obtained through the

use of nature (for subsistence, recreation and tourism, utilization of genetic diversity) and the values of *ecosystem functions* (such as climate regulation and the maintenance of matter cycles). The value of biological diversity also includes the *option values* for future generations and *existence values* (WBGU, 1996).

Although biodiversity is of great global significance, it is currently undergoing a dramatic decline (habitat destruction, species extinction, genetic erosion). Moreover, climate change may cause rapid shifts in climatic zones, which may override the adaptability of ecosystems. This in turn could further accelerate the destruction of habitats and species. For this reason, loss of biodiversity must be regarded as one of the core problems of global change (WCMC, 1992; WBGU, 1993 and 1996).

3.4.2
Major Contributions by German Biodiversity Research

The organization of *Ecosystem Research Centers* (Kiel, Göttingen, Bayreuth) within TERN (Terrestrial Ecosystem Research Network) and interdisciplinary research concepts (e.g. those implemented at ZALF in Müncheberg, PIK in Potsdam and UFZ in Leipzig-Halle) are promising approaches for biodiversity research. The ecosystem research centers are attached to universities, well staffed and equipped, and possess good facilities for conducting research on global environmental change (research on specific ecosystem functions relating to global matter cycles and climate change is dealt with in *Section B 3.1*). Successful cooperation between German Research Centers and universities in the field of soil ecology is also outstanding (Wissenschaftsrat, 1994). However, this research is primarily focused on the natural sciences, with socioeconomic issues receiving little attention.

German *research into forest decline* also provides inputs to biodiversity research. It seeks to identify response strategies for stopping the die back of forests that has been increasingly apparent since the mid-1970s and which is brought about by a complex interrelationship of causes and effects in which air pollutants play a key role (BMFT, 1990). By applying an interdisciplinary and interinstitutional approach, forest decline research has led to some important results (though by no means conclusive, nor the simple results originally expected) and has developed specific strategy response options (Wissenschaftsrat, 1994). The knowledge acquired through forest decline research and the methodological progress achieved in field and laboratory investigations could be applied in other regions besides the temperate zones to determine the damage caused to natural ecosystems by airborne pollutants. Many newly-industrializing and developing countries see themselves confronted, today or in the future, with similar and in some cases more severe air pollution than in central Europe. The Institute for World Forestry in Hamburg coordinates data collection and evaluation for the European forest status report; its other activities include the generation of concepts, indicators and criteria for sustainable forest management.

It is difficult to pinpoint any outstanding German contributions to biodiversity research besides those mentioned above. Research in this field is still predominantly characterized by individual projects at local or regional level, whose strengths are essentially based on their application within a local context (such as species and biotope inventories, Red Lists and management of cultivated lands). Taxonomy, which provides the main foundation for biodiversity research, hardly plays a role at all in Germany (Ziegler et al., 1996). Ecological research relating to conservation exhibits certain deficits, as does conservation-oriented research for the developing countries and eastern Europe (Wissenschaftsrat, 1994); to date there have been virtually no genuinely interdisciplinary research programs in this field. One of the exceptions is the Tropical Ecology Support Program (*Tropenökologisches Begleitprogramm*) of the GTZ, which also promotes research on ecological aspects within the framework of development cooperation (GTZ, 1995).

Analyzing the activities of the German Research Foundation reveals that the requirements for modern biodiversity research as defined above is met only by a few projects (e.g. in the Mechanisms for Maintaining Tropical Diversity priority research program at the University of Würzburg). Collaborative research centers are not really involved in biodiversity research as such. With the exception of a graduate college at the University of Mainz, research groups and innovation colleges have not been used as instruments in this field up to now either (Ziegler et al., 1996).

The UBA Environmental Research Catalog (UFOKAT) lists, under various headings, a number of projects involving biodiversity research (UBA, 1992). In most cases, however, these are small, locally or regionally confined projects lacking an interdisciplinary approach or transdisciplinary coordination.

Germany is relatively underdeveloped compared to most Anglo-American countries as regards the application of molecular biology methods for assessment and monitoring of biodiversity, for biodiversity conservation, for biodiversity prospecting, for rehabilitating degraded ecosystems, or in the field of biosafety research. The gaps in globally oriented biodiversity research mean that sound arguments will be lacking in any discussions that are necessary prior to making a decision on conservation measures, thus resulting in inadequate acceptance and enforcement of conservation (Council for Conservation and Landscape Management at the German Ministry of the Environment, 1995 a and b).

New initiatives for supporting international biodiversity research have yet to emerge in Germany. In other countries, by contrast, numerous organizations already pointed to the need for greater promotion of modern taxonomy and research on tropical biodiversity several years ago. In Great Britain, for instance, the Darwin Initiative was established with the aim of training scientists in developing countries, while the program Partnerships for Enhancing Expertise in Taxonomy (PEET) was set up in the USA to counteract the loss of such taxonomic expertise. Lack of knowledge in this area is seen as a limiting factor for biodiversity research (Stork and Samways, 1995). Both countries have a significant edge over Germany in the field of modern taxonomy, where molecular biology methods are also used. It is therefore fair to say that German taxonomic biodiversity research only has a leading international position with respect to the systematic identification of microorganisms and their functional diversity.

3.4.3
Integration of German Biodiversity Research in International Programs

The UNESCO "Man and the Biosphere" program (MAB) is not primarily a research program, although research and monitoring activities are carried out and coordinated within its framework (*see Section B 1.4*). MAB is an attempt to create a synthesis between environmentally sound land use (taking economic and social aspects into account) and conservation. Program activities in Germany include the designation of 12 biosphere reserves covering about 3.3% of German territory. Some ecosystem types are still not included, however, such as urban and industrial landscapes and intensively cultivated agricultural land. Priority will have to be given in future to these types of ecosystem when new biosphere reserves are designated (Erdmann and Nauber, 1995). The German contribution to MAB focuses primarily on the creation of ecosystem centers and biosphere reserves serving national interests (Wissenschaftsrat, 1994), although the German National MAB Committee is endeavoring to set up a Monitoring Program in Biosphere Reserves (BRIM). The Secretariat of the German National MAB Committee conducted a survey among European biosphere reserves in 1993/1994 in order to determine the potential for monitoring and research activities in permanent observation areas. Following Germany's lead, regional cooperation in Europe has been institutionalized in the form of EUROMAB.

There are few internationally oriented and integrated activities for biodiversity research. One that deserves mention here is the Diversitas program jointly established by IUBS, SCOPE and UNESCO for the promotion and coordination of biodiversity research, which aims at providing information, developing predictive models on the status and sustainable use of biodiversity, and capacity building (Diversitas, 1995). Several countries have started to translate this approach into national research policy (Stork and Samways, 1995). Germany is not involved in this program to any significant extent.

The Global Biodiversity Assessment (GBA) was commissioned by UNEP in order to obtain a comprehensive review of current knowledge in the field of biodiversity (Heywood and Watson, 1995). More than 1100 scientists from 80 countries cooperated in this key project on the status of biodiversity research. The fact that only six German experts took part, none of them in a major capacity (as coordinator or member of the steering group), is characteristic of the lack of international involvement on the part of German biodiversity research. Moreover, there is no major German involvement in the design and implementation of other international research activities (such as Systematics Agenda 2000, BioNET; *see* Stork and Samways, 1995). Consequently, the number of biodiversity research publications by German scientists that are cited in the international literature is low.

3.4.4
German Biodiversity Research – Research Needs Concerning Global Change

3.4.4.1
Scientific Issues

The wide range of disciplines that are implicitly or facultatively involved in biodiversity research makes it inappropriate to structure research needs according to disciplines. In the following, we dispense with such a formal approach and focus instead on global issues (as is done in the Global Biodiversity Assessment, Heywood and Baste, 1995).

BIOLOGICAL ASPECTS
Inventorying, Classification and Monitoring
Despite the fact that taxonomy dates back 200 years, only a fraction of the total species diversity has

been described so far. One very important contribution that a biodiversity research program could make would be to develop criteria and methods for studying diversity at the level of organisms and populations (Solbrig, 1991). International consensus regarding the methods and priorities for systematic inventorying of species in the world does not yet exist, and would need to be established. There is inadequate harmonization of existing initiatives. Starting from a systematic and comprehensive inventory of biodiversity, it should be possible to generate predictive classifications reflecting the "history of life" and to store this knowledge in a database that must be accessible to all countries and researchers (Diversitas, 1995).

The data we possess on species is not organized in an optimal way. For example, there is no *global master list* of species known to humanity. Collections and descriptions of species are widely strewn, are found largely outside the countries of origin and are difficult to localize. Adequate infrastructures and research capacities for systematic identification and description have yet to be created worldwide, and existing obstacles to the characterization and understanding of biodiversity have yet to be removed.

The Dynamics of Biodiversity
Fundamental research is necessary at different levels if we are to advance our understanding of biodiversity dynamics. It is essential to understand the underlying genetic processes – particularly the units and mechanisms of selection – in order to explain the relationship between diversification and the extinction of populations or species. Many basic issues relating to organisms and species have still not been clarified to any adequate extent:
- Agreement on the definition of the species concept.
- Origin, dynamics and assessment of species diversity.
- Relations between species diversity and ecosystem structure (concept of keystone species, diversity versus stability, minimal required diversity, redundancy, etc.).
- Interrelationships between ecosystem structure and ecosystem function.
- Human influences on biodiversity dynamics.

The lack of empirical data on the diversity of organisms in many natural ecosystems makes it much more difficult to study species diversity among biological communities. Elaborating hypotheses for certain taxonomic groups in selected geographical regions and then applying these hypotheses to other groups and ecosystems is the method often used, but this is problematic to the extent that it is not known how representative a group of organisms is in each case (Solbrig, 1991).

VALUATION, CONSERVATION AND SUSTAINABLE
USE OF BIODIVERSITY

The interactions between biodiversity and human societies raise questions concerning the anthropogenic influences on biological diversity, the sustainable use of biodiversity and the fair and equitable sharing of the monetary benefits resulting from the use of biodiversity resources (*see* Article 1 of the Biodiversity Convention). One key complex of issues involves the economic values attached to biological diversity by society and individuals, which range from the philosophical and social basis for the conservation and sustainable use of biodiversity, to people's relationship to and experience of nature. Economic research aspects include the basic question of how to value biological diversity. Specific issues in this connection are the categorization of economic values, problems related to the internalization of these values, and the related debate on potential economic instruments (*see Sections B 3.6.4.1 and B 3.8.2.1*).

Knowledge derived from biodiversity research must be understandable for the political and administrative spheres, and must provide for concrete action. Priority should be attached in this context to the development of strategies for the conservation and sustainable use of biodiversity, whereby fundamental questions have yet to be clarified:

* Should priority be given to the condition of ecosystems, or to processes?
* Which species or communities require the greatest protection?
* How can the spatial requirements of biotopes be determined?
* How should one respond to the effects of climate change on ecological systems, given the low level of knowledge about expected regional impacts?

Before policies can be implemented, it is necessary to develop improved methodologies, instruments and communication for inventorying, monitoring, risk assessment and management of biodiversity. Research on standards and indicators (both generalized and regional in nature) as well as on procedures for assessing the condition of and changes in ecosystems is essential in this connection. These research efforts should also aim at integrating the conservation and sustainable use of biodiversity into the various levels of planning.

Sustainable use of biological diversity is a goal that demands research relating to the conservation of genetic diversity of useful plants and animals, the development of integrated uses (agroforestry, etc.), and biodiversity prospecting (*see* WBGU 1996 for detailed research recommendations). Special attention should be given to the consequences of different land-use regimes in agriculture, forestry and fisheries when investigating anthropogenic effects on biodiversity (landscape diversity). The decline of genetic diversity among agricultural plants and livestock (genetic erosion) is a result of narrowing the number of species used and replacing traditional and locally adapted crop varieties and landraces with manipulated breeds and with the large-scale use of high-yielding crop varieties and livestock breeds.

Gene banks (e.g. the Institute for Plant Genetics and Crop Plant Research in Gatersleben) are of major importance for research on and maintenance of genetic diversity. There are already some international initiatives to combat the genetic erosion of cultivated species through the exploration, collection, documentation and conservation of plant genetic resources (the International Undertaking on Plant Genetic Resources, FAO). The 4th International Technical Conference on Plant Genetic Resources (Leipzig, June 1996) detailed the current status in this field and adopted a Global Plan of Action for the conservation and sustainable utilization of plant genetic resources. The German national status report submitted to the conference also addresses the need for research on plant genetic resources (BML, 1996).

Major gaps still exist with regard to the social scientific aspects of biodiversity. A fundamental issue concerns how the underlying economic and social driving forces can be defined, and altered with respect to their destructive effects for the conservation of biodiversity. This involves investigating, for example, the impacts of international trade on biodiversity, the design and implementation of international conventions and the relevant economic instruments and regulatory frameworks.

Another research gap concerns the question of how to ensure participation of indigenous and local communities in the sustainable use and conservation of biological resources, as required by AGENDA 21. Much work remains to be done in connection with capacity building in order to clarify how, in the field of biodiversity, the transfer and use of data, methodologies, financial resources and technologies can best be organized.

3.4.4.2
Strategy for Future Biodiversity Research

As explained above, the global aspects of biodiversity research are of great importance. Nevertheless, regional and national research approaches may also have global significance, in that the loss of biological diversity is primarily caused through the accumulation of national and regional factors (such as poverty, nutrient inputs, overexploitation, clear cutting, land-use changes, erosion and desertification). Global problems such as the anthropogenic greenhouse effect and enhanced UV-B radiation do not cause biodiversity loss to the same extent as yet, but need to be researched intensively on account of their potential future impacts. Priorities for future biodiversity research in Germany therefore include:

- *Conduct of and participation in international taxonomic projects for the inventorying of species* (analogous to the projects managed by Diversitas or Systematics Agenda 2000).

- *Conduct of and participation in a global biogeographical survey of biodiversity*. Focal points of research in this area would include, in particular, the changes in biodiversity resulting from anthropogenic disturbances and the operation of anthropogenic factors within ecosystems, which would require the development of a methodology for interregional comparisons of biodiversity (Solbrig, 1991). Gradients of species diversity between coastal and mountainous regions, between wet and dry regions, warm and cold climates or between freshwater and saltwater for the lower latitudes should be described and compared with the relevant gradients from middle latitudes.

- *Research on the impacts of global change on biodiversity and ecosystems*. Some topics and keywords include: reactions of ecosystems to contamination and/or to changes in climate and water resources (*see Section B 3.1*); significance of species diversity for the development of stable landscapes; use of monitoring and modeling for describing the local, regional and global development of ecosystems.

The sheer magnitude of the research needs listed in *Section B 3.4.4.1* is a clear illustration that the tasks at hand cannot be accomplished unless there is international coordination on the defining of priorities and the division of labor. In the following, we specify four areas in which there should be intensive involvement on the part of the German research community:

1. Some of the weaknesses displayed by German biodiversity research stem from the severe decline in the fields of organism biology, biological systematics and taxonomy. These disciplines are wrongly considered to be "antiquated". Experts have been saying for years (Henle and Kaule, 1992; Sukopp, 1992) that modern taxonomy must play a central role in education and research, because it provides the core reference system for all other areas of biodiversity research (Bisby, 1995). Hubert Markl, President of the Max Planck Society, recently drew attention to this deficit, saying that "without the active contribution of lively and productive biotaxonomic research, especially into the organism inventories of the tropics and subtropics and the oceans at all latitudes, it will be impossible to acquire the ecological knowledge necessary to achieve the global management of the biosphere – both management of it to provide benefits for ourselves as well as our protective self-restriction towards ecological communities – in such a way that we succeed in establishing the long-term sustainable coexistence of nature and humanity on which our future depends." (Markl, 1995). The DFG (1992) and the *Wissenschaftsrat* (German Science Council) (1994) have also demanded intensified efforts in the neglected field of taxonomy. The DFG criticizes the lack of discussion between systematics experts and molecular biologists, which has led to the wasting of opportunities in the modern field of biochemical characterization. Major hopes are placed in strategies for the participative use of natural, biotechnologically useful substances for in-situ conservation of biological diversity (biodiversity prospecting). Accordingly, greater attention should be devoted to the field of chemical ecology and research on natural products, with participation by industry and the social sciences as well.

2. *Population biology* has also gained significance internationally, partly in connection with advances in molecular biology, but also through behavioral and socio-ecological concepts. However, both the population ecology and population genetics of wild species are severely under-developed in Germany compared to other countries. A key task is therefore to establish population ecology research geared to nature conservation and to develop it to international standards (Kaule and Henle, 1992). Furthermore, new methodologies in the field of population genetics now provide key concepts of evolution with specific data and experimental results.

3. Another key area of biodiversity research in which Germany should become more involved is *biodiversity economics*. This sub-sector of environmental economics has undergone rapid expansion in recent years, partly in response to the Biodiversity Convention and the discussion on the value and valuation of biological resources, and has led to many publications and contributions. A number of German researchers are now working in this field, but there is still a general lack of interdisciplinary or nationally coordinated research.

4. Another field that has gained in importance through the more recent international conventions is that of convention-related research, i.e. research on the development and implementation of the Convention on Biological Diversity (e.g. the draft Protocol on biosafety) and other conventions relevant for biodiversity (such as CITES, Ramsar and the Berne Convention). Contributions to this field are likely to originate from the

political science domain primarily, but must also acquire an interdisciplinary dimension through networked links to the life sciences. The issue is where this research should be located – in existing institutions (the Federal Agency for Nature Conservation, for example) or at universities (*see next section*).

3.4.4.3
Organization and Structure of Biodiversity Research

Given the importance of biodiversity research, there is an urgent need in Germany to improve the organization and structure of research in this field and to enable better integration within the international research effort. The basic principle governing any such improvements is that biodiversity research requires the promotion of multi-layered interdisciplinary and inter-institutional cooperation. In view of the direct political relevance of biodiversity, a further important aspect is that the research community become more aware of its role and responsibility for the management and dissemination of its results as the basis for political action.

This calls for an unequivocal prioritization of biodiversity issues within research policy. The Council appeals for the establishment of an independent government funding program for biodiversity research as a key area with an enormous range of applications and economic potential. One should examine the extent to which greater networking could be achieved between the various biodiversity research projects and facilities already receiving support through BMBF research funding programs (the relevant institutes would include, e.g. the Hans Knöll Institute for Natural Products Research or the agricultural research centers in the Biotechnology program, e.g. the TERN centers in the Environmental Research and Climate Research domain, e.g. the AWI in the Marine Research and Marine Technology, Polar Research field, the support concept for Renewable Raw Materials in Energy Research and Energy Technologies; other support measures relating to biodiversity exist in the other fields as well). A Biodiversity government funding program should also involve participation on the part of the private sector.

Analogous to the existing Tropical Ecology government funding priority, the establishment of additional DFG funding priorities on biodiversity research (the German Science Council has recommended the establishment of funding priorities for applied ecology and nature conservation) would be a response adequate to the importance of this research field and would close the gap between fundamental research and practice-related data collection, as well as that between the life sciences, on the one hand, and law, economics and the social sciences, on the other. The BMBF's efforts in the field of hydrology research to bring together fundamental research groups at the universities and the more applications-oriented working groups within the regional authorities could be models for biodiversity research to follow.

A plausible response would be to establish separate degree courses dedicated to the in-depth study of biodiversity, thus complying with recommendations to integrate nature conservation research (including biotechnological content) as a component part of thorough education in the life sciences. At the same time, the fundamental importance of holistic biology would acquire new relevance by embracing modern methodologies in the field of molecular biology. Above all, such degrees or specialized courses would attach commensurate importance to the science of biodiversity within the educational sector.

Biodiversity research initiatives at European level have already been proposed (Heywood, 1993); these initiatives should be taken up in Germany, developed programatically and submitted to the European Council of Ministers for further action. Thought should be given not only to a separate promotion of biodiversity research within the EU, but also to the integration of biodiversity in the support frameworks for central and eastern European countries (TEMPUS, PHARE, etc.). The EUREKA Programme would offer a good Europe-wide framework for international and interdisciplinary technology-oriented biodiversity research.

Many fields of biodiversity research demand international frameworks and cooperation networks, so one key focus for the future strategy of German biodiversity research should be to establish an *international presence*. German researchers should participate more intensively in international biodiversity initiatives such as Diversitas, BioNET International or Systematics Agenda 2000, and should generate new initiatives for joint international projects.

3.5
Population, Migration and Urbanization Research

3.5.1
Relevance of Population Size, Migration and Urbanization to Global Change

The Council has repeatedly drawn attention (WBGU, 1993, 1995a and 1996) to the fact that the growth and distribution of the world's population are issues of central relevance for analyzing and coping with global environmental problems. In many states, high population growth is both a cause and a consequence of poverty and environmental destruction. Other reasons for the high birth rate include, as ever, the widespread discrimination against women in many social domains, inadequate access to basic medical care, and lack of education. As a result of population growth and mounting poverty, near-natural and often marginal locations are transformed into agricultural land. This trend is enhanced in rural areas through the diminishing factor productivity and declining soil fertility of the land brought about by inappropriate land-use practices. Migration to the rain forests of Brazil or Indonesia, the transitional zones towards the southern Sahara, or to mountain slopes in the Andes and the Himalayas leads to environmental damage that not only destroys local ecosystems, but also has regional and global impacts.

Moreover, the destruction of the natural environment is one of the major triggers – besides civil wars and natural disasters – of large-scale migration from rural to urban areas in the developing countries (UNDP, 1992; Hauser, 1990 and 1991). Africa, India, Southeast Asia and Brazil are among the worst affected by this trend. Towards the end of this century, half of the world population will be living in cities, and the figure will rise by the year 2025 to two thirds. Inadequate housing and increasing homelessness are alarming signals of the problems faced by cities and represent a serious threat to the health and security of their inhabitants.

This crisis is exacerbated by the growing numbers of refugees, the lack of employment, the growth of slum areas, widening disparities between rich and poor, dilapidation of urban buildings and infrastructure, increasing air and water pollution and the increasing vulnerability of urban populations to natural and human-induced crises (*see Urban Sprawl and Favela Syndromes in Section C 2.2.2*). These problems are particularly severe in the *megacities* of the world (WRI et al., 1996).

Intranational flows of refugees can ultimately develop into international migration flows that will also affect industrialized "immigration countries", which means that the problems of population growth, migration and urbanization will necessitate response strategies on the part of the developed countries.

3.5.2
Major Contributions by German Research in the Field of Population, Migration and Urbanization

German global change research on population, migration and urbanization issues has been focusing for only a few years on the interactions between poverty, population growth, migration, urbanization and environmental destruction. It is exceedingly difficult to pinpoint areas where Germany is making an outstanding contribution, due to the inadequate networking of research establishments in this field, the lack of transparency and the dilution of research activities across different university institutions and public-sector research facilities.

A number of disciplines – social anthropology, geography, regional planning, sociology and economic science in particular – are subjecting these issues to intensive analysis in the form of empirical case studies and theoretical work. Although these individual disciplines address specific aspects of the problem and could provide key elements for a more global perspective, their current research activities lack the essential global focus. *Systematic networking* of the disparate research activities in Germany, and *integration into the international effort* in general are needed as a matter of urgency, if the sought-after synergy effects and greater transparency among researchers is to be achieved.

3.5.3
Integration of German Population, Migration and Urbanization Research in International Programs

Global change research in the field of population, migration and urbanization plays a much lesser role at national than at international level. Involvement in international research projects should be organized more intensively through the International Human Dimensions of Global Environmental Change Program (IHDP) (*see Section B 1.3*).

One of the main foci of the German global change research program since 1992 has been the study of relations between people, society and the environment, and the implementation of the sustainable development concept. These aspects are addressed, for example, in the DFG priority program People and Global Environmental Change – Social and Behavioral Dimensions.

At the European level, German population research is integrated within the EU's Fourth Framework Programme for research and technological development and demonstration (1994-98) (*see Box 5*). Issues pertaining to population trends, urbanization and migration are addressed in the specific program for Targeted Socioeconomic Research (TSER), which has been allocated a budget of approximately US$ 23 million.

To improve cooperation and exchange of experience, the World Bank, UNDP and HABITAT (UNCHS) set up the Urban Management Programme (UMP), which Germany also supports. The program aims to collect and develop experience in the field of urban environmental management. The UMP has been in operation since 1986, has a budget of approximately US$ 25 million, and is the biggest multilateral urban development program in existence. The priorities for the second project phase (1992-1996) are urban finance and administration, infrastructure development, land management, urban environmental management and the eradication of poverty.

The Maastricht Health Research Institute for Prevention and Care (HEALTH) was set up at the University of Limburg in 1995. The WHO Collaborating Centre for Research on Healthy Cities is an integral part of the latter. One of the main research thrusts concerns the political level of sustainable urban development, and German participants of the associated Healthy City Project (HCP) are Dresden, Frankfurt am Main and Hamburg.

3.5.4
German Research in the Field of Population, Migration and Urbanization – Research Needs Concerning Global Change

The factors causing population growth, migration, urbanization and the resultant environmental changes cannot be viewed in isolation from each other. The Council has repeatedly drawn attention to the complex web of causes and effects that researchers must take into consideration in their approaches and methodologies. This relates in particular to the following systemic linkages.

3.5.4.1
Urban-Rural Relations

The quality of intranational urban-rural relations is characterized worldwide by simultaneous rural exodus and remigration to villages (*see Rural Exodus Syndrome in Section C 2.2.1*). Both aspects deserve attention.

The governments of developing countries, faced with the pressure of rapid urbanization, are unlikely to reduce the urban bias (the almost exclusive concentration of financial resources on cities) that has shaped their national settlement policies until now. Instead, the disadvantages suffered by rural areas will be further exacerbated, and the willingness to migrate will increase. Although there is a general recognition that the underdevelopment of rural areas contributes more towards *rural-urban migration* than the attractivity of the cities, those regional development programs that have been set up have not slowed rural exodus to any perceptible extent.

One consequence that is already visible in Africa and in some parts of South and Southeast Asia, is the reshaping of urban-rural relations. This involves the replacement of the traditional cliché of rural exodus and the exploitation of rural areas by urban agglomerations by a new understanding in which migrations are interpreted as exchanges (Kreibich, 1992). The more difficult it becomes to survive and to find economic and social security in the cities, the greater the attractivity of rural areas, at least in those countries that still have unexploited reserves of land and productivity (especially in Africa). The decision to migrate must therefore be interpreted in the context of trends within the rural economic system as a whole, and above all at individual household level, in order, for example, to explain the growing proportion of women in the urban influx, or the high remittances transferred by migrants.

The "circulation" of large population groups between urban and rural areas, the effects this has on transformation processes within rural society, and the growing links between urban commodity economies and rural subsistence economies are now addressed by some novel research approaches – at the universities of Berlin, Bielefeld and Freiburg, for example. An important result of this change of perspective is a reinterpretation of urban-rural capital flows; contrary to traditional analyses, survival in large cities is actually determined to an increasing degree by the transfer of money and goods from rural areas. Some researchers even reverse the classical model of push and pull factors (Unwin and Potter, 1989), having discovered a growing willingness to remigrate to the country on account of deteriorating living conditions in large cities and the new attractivity of rural areas and smaller towns. This leads to the following research needs:

- Relations between cities and their rural surroundings must be re-investigated and re-assessed with respect to the factors mentioned above, in order to embrace the growing complementarity of migration flows and the exchange dimension this implies. At the same time, attention should be focused on the growing importance of transfers from rural to urban areas for survival in urban agglomerations. The specific locations (e.g. regional markets), the actors (e.g. labor migrants, traders, transporters), the typical mechanisms (e.g. seasonal migration, remigration on retirement, social networks) and the systematics of urban-rural exchange must all be subjected to more precise analysis.

- New approaches to *rural development*, aiming at generating alternatives to agricultural production and improving infrastructure, should be developed through cooperation with institutions that are practically involved in this field (e.g. the GTZ). Such activities should take into account the changes that have occurred in the overall system of rural economies and the systemic linkages with urban areas.

3.5.4.2
The Individual's Decision to Migrate

The migration of individual household members to cities is one of several options for ensuring survival that are open to rural households. Alternatives include the cultivation of new land, the intensification of agricultural production or switching to field crops with a high yield or high market value. Some of these alternatives provide households with temporary relief, but threaten or even destroy their life-support systems over the long term. The newly-industrializing countries provide ample evidence to corroborate the thesis that increasing agricultural productivity is an insufficient basis on which to reduce migration from rural areas; it must be coupled instead to a general improvement in living conditions in rural areas (*see Rural Exodus Syndrome, Section C 2.2.1*), above all through the establishment of a labor market outside the agricultural sector. Most developing and newly-industrializing countries have great difficulty achieving this, due to the dominance of large towns and cities and inefficient regional planning. Instead of upward social mobility, migrants are only able to find a modest livelihood – often as members of "marginalized" social groups. The decision to migrate is thus a very difficult one, and depends on the range of options open to those involved; too little is known as yet

about which options ultimately trigger migration. More research is therefore needed in the following areas:

- The individual's perception of the situation faced by the household must be ascertained before one can postulate links between migration and environmental factors; special attention must be given to sociocultural (lifeworld) factors as well.

- The decision to migrate must be modeled in a realistic manner. Special focus must be directed to the specific family or household situation, the temporal perspective (circular migration) and the economic significance of migration (creation of agrarian capital). Traditional "flow analysis" of the migration process must be replaced or enhanced by a form of "migration system research" that examines the origin, direction and duration of migration in the context of different causal complexes.

3.5.4.3
Food Security

One of the greatest challenges that global change poses to humanity is how to achieve universal food security in the face of constant and rapid growth of the world population. Undernourishment, malnutrition and famine are major causes of migration – in urban and rural areas to equal extents. Ensuring humanity's freedom from hunger involves not only food *production* (carrying capacity, Green Revolution, genetic engineering) and *distribution* (world food trade, food aid, agricultural policy), but also the *institutional framework* (property rights, empowerment of people, equitable participation on the part of social groups). For all the successes of the Green Revolution, in which per-capita food production was greatly increased, the number of undernourished is growing in most developing countries, especially in Latin America, the Middle East and Africa (*see Green Revolution Syndrome, Section C 2.2.2*).

At the same time, the ecological problems associated with monocultures of high-yielding varieties susceptible to disease and pests are becoming increasingly apparent. One study on agricultural carrying capacity carried out by the FAO revealed that agricultural production in many countries in sub-Saharan Africa could already be increased sustainably through improvements that are based on traditional production methods and do not involve imported equipment. The following research activities are needed in order to shed light on these problems:

- The concept of (agricultural) carrying capacity and the methods for calculating carrying capacity must be refined to cover the dimensions of technological factors (productivity) and environmental impacts; greater distinctions must be made between different agricultural systems and land management methods.

- A more detailed analysis is needed of other determinants of malnutrition, undernourishment and famine besides those on the supply side.
- Given impending climate changes, it is essential to improve harvest forecasting methods and early-warning systems (e.g. against drought).

3.5.4.4
The Informal Sector: Labor Market and Ensuring a Livelihood

The number of urban poor is constantly rising in developing countries (as is the wealth of the small upper classes). At least half the population of the new (mega)cities live in marginal settlements, partly in inner-city slums and shanty towns, but above all on the periphery of agglomerations, where almost every building has been erected without official planning and approval (squatters) and where even the most basic technical and sanitary infrastructure is absent (*see Favela Syndrome, Section C 2.2.1*).

Twenty years ago, the research community could hardly have imagined that the agglomerations in the developing countries could grow so rapidly under conditions of extreme poverty without a severe increase in social and political conflicts (Mertins, 1992). One of the keys to understanding this unexpected development lies in the concept of the informal sector, which has since been subjected to continuous refinement and empirical study (e.g. at the universities of Bielefeld and Freiburg). As a rule, the two economic sectors must be analyzed in relation to each other, given that the formal sector could no longer exist in many cases without the output of the informal sector. Thus there is still a need for research in this area:

- The informal sector plays a central role for the urban poor in maintaining a minimum of social security. However, the extent to which it represents a separate development potential has been subjected to little investigation so far. This research aspect seems all the more important given the failure of traditional modernization strategies.

3.5.4.5
The Informal Sector: Human Settlements Development

Compared to the informal labor market, there has been relatively little research on *informally constructed cities*. It is not possible to understand how and why agglomerations in the developing countries

did not collapse long ago unless we have knowledge about the control mechanisms behind the informal construction of homes and settlements. How could a city like Dar-es-Salaam, for example, whose population expanded within twenty years from a few hundred thousand to three million, grow to such an extent without a land registry and effective urban planning, without epidemics and major conflicts?

A characteristic feature of new settlement structure is the "city of large villages" which forms along rural-urban link roads, on marginal land (flood-prone areas and mountain slopes), but especially around village settlements in the urban periphery, where land is still available.

The respective legislation governing land tenure similarly determines the spatial form that settlements take. New settlements are not connected to the basic utilities that probably exist in core areas of the city, and in many cases do not have weatherproof access roads. Although these peri-urban settlements are inadequately integrated into local hierarchies, they nevertheless exhibit elements of urban economy and society: markets, retail shops, and the production of goods and services.

In the medium term, there is a risk that fewer and fewer people will be able to develop informal settlement structures without external assistance. They will come up against barriers that are insurmountable with the means available – the abandonment of agricultural land, worsening sewage disposal problems as a result of increasing settlement density, rising traffic levels, and intense competition for jobs in the informal sector of a stagnating national and/or regional economy with decreasing links to national and international markets. The research needs implied by these interlinkages are concentrated on the following topics:

- Informal construction of housing (and commercial buildings) and *informal settlements development* must be studied in terms of their systemic interrelationships. Knowledge of the mechanisms that operate is indispensable, not only to understand the formation and functional operation of unplanned large-scale agglomerations, but also to find starting points for effective settlements development measures where resources are extremely limited.
- Too little is known about the internal limits (potential for regulation, costs) and external limits (environmental impacts) to informal settlements development. Suitable models and indicator systems for predicting the maximum absorptive capacity and/or capacity for survival of urban agglomerations have yet to be developed.
- The systematic articulation between the formal and informal sectors should be studied more in-

tensively in order to find ways to achieve sustainable urban development.

3.5.4.6
International Migration Flows

If the problems relating to urban-rural relations, the informal sector and intranational migration flows described above worsen, as has been the case in recent years, it must be assumed that the push factors driving international migrations will intensify. According to estimates of the United Nations Population Fund (UNFPA), approximately 100 million people (around 1.7% of the world population) were living outside their country of birth in 1995. About 14.4 million of these were refugees and asylum-seekers, while another 13 million were in refugee-like situations. The number of internally displaced persons, i.e. refugees who flee to other parts of their country to escape internal unrest, violence, drought and environmental disasters, is in the order of 23 million.

Potential immigration countries, Germany among them, are likely to face serious problems as a result of the expected increase in migration pressure. To respond to this situation, research in the following areas is needed:

- Any migration policy that is not founded on reliable forecasts of the number and origin of future migrants is liable to result in disastrous mistakes being made. There should be a shift of research focus towards identifying the regions in which international migration flows originate, in order to respond to foreseeable trends in a timely and controlled manner on the basis of such data.
- Systematic knowledge about the specific motivations underlying migration is essential if we are to recognize and understand global changes, and hence draw accurate conclusions regarding the future direction and extent of international migration flows.

3.5.4.7
Megacities Within the Global Network System

Between 1980 and the year 2000, the percentage of urban dwellers in the industrialized countries will increase only slightly to around 75%, whereas the figure for the developing countries will rise by about 10% to almost 40%, and to an estimated 53% by the year 2020 (UN, 1995). In Latin America, the proportion of the population living in urban areas will exceed that in the industrialized world by the year 2000. Africa, which is still considered a "rural" continent, already has a higher level of urbanization than Asia,

as well as the world's highest urbanization rate. By the year 2020, almost one person in two in the developing countries, a total of 2.2 billion, will be living in cities with a million or more inhabitants. According to UN estimates, Africa will by then have more than 30 cities with at least 4 million inhabitants, and these will be integrated into global flows of capital, information and goods within the context of advancing international division of labor. The question is whether the global networking of megacities and the concomitant effects on human welfare will boost the development of surrounding rural areas, or whether these urban agglomerations will develop into islands with few linkages to their rural surroundings. From this we derive the following recommendations for research:

- Our knowledge about large-scale agglomerations in developing countries and their integration into the global system is still patchy. Much work has to be done to identify phenotypes and their specific interactions with the regional economic, social, political and cultural context.

3.5.4.8
Education

About one fifth of German Official Development Assistance, or US$ 1 billion, was allocated to the education sector in 1993. Partly as a result of such educational aid, the number of illiterates has declined worldwide since 1980 (to around 885 million in 1995). Nevertheless, the educational gaps and disparities in knowledge between industrialized and developing countries, and among the latter, have widened further. This trend threatens not only economic development in the developing world; an inadequate level of basic education is also a threat to population control, the conservation of the environment, the promotion of democratization, the maintenance of peace and the realization of human rights (BMZ, 1995).

Many factors prevent poor households from using the educational system and obtaining the benefits it provides. If children must work within the family or in external employment, sending them to school involves prohibitive costs. The benefits of basic-level education are then enjoyed by better-off families only. Another major problem is that many countries fail to provide women with equal and adequate access to education (WBGU, 1996). From this we derive the following research needs:

- A minimum level of education, especially for women, has a positive influence on the overall development of society. In many cases, educational opportunities are not taken advantage of due to language barriers. Investigations should therefore focus on the extent to which greater decentralization of the educational system may prove a better way of providing first-language education.

- Until now, research into the quality of teaching has mainly involved comparative studies that concentrate on cognitive learning in mathematical and natural science disciplines. By contrast, studies on the practicality of existing objectives and programs are rare. If scarce educational resources are to be exploited effectively, research should also focus on the available options for optimizing the provision of education.

3.5.4.9
The Social Position of Women

The United Nations World Conference on Women and Development held in Peking in 1995 highlighted the fact that men are generally better situated than women in most developing countries as a result of sociocultural norms. The burdens to which women are exposed are all the more evident if one compares the number of live births that women in different countries have. The fertility rate in many African countries is still 8 or more, i.e. many women are pregnant or nursing newborns for more than a third of their adult lives (Dasgupta, 1995). In most developing countries, complications during pregnancy are still the most common cause of death among women of childbearing age. The WBGU has repeatedly emphasized the key function of women for the positive overall development of societies (WBGU, 1995a and 1996) and identifies the following areas where research is needed:

- In view of the incongruence between the skills and responsibilities of women, on the one hand, and their real opportunities in society, on the other, research activities should be concentrated, *inter alia*, on improving the legal position of women, especially in respect of obtaining credit, land or productive means.
- Women must be empowered in accordance with their interests, needs and sociocultural self-understanding. Above all, the specific situation of women in developing countries must be subjected to more detailed analysis.

3.5.4.10
Health

Poverty and underdevelopment are the principal causes of health hazards in developing countries. The consequences of poor health are lost working days,

premature invalidity and hence enormous costs to the national economy (BMZ, 1995).

Although antibiotics, the increasing use of vaccinations and anti-malaria campaigns have definitely led to increased life expectancy in the developing countries (UNDP, 1992), about 1 billion people still suffer from acute vitamin or mineral deficiency.

According to the latest WHO estimates, 14 million people are infected with HIV, a figure that will probably rise by the end of the decade to somewhere between 30 and 40 million. The German Foundation for International Development (DSE) has analyzed the disastrous consequences that the spread of Aids is having, particularly since it mostly afflicts young adults in key economic positions, especially in rural areas. Staff of the USAID Aids Program fear that Asia is likely to have more new cases of HIV and Aids in the next few years than Africa. By the year 2000, according to UN estimates, the number will increase by one million annually (UN, 1994).

- In addition to the acute need for action to tackle these problems, there is an urgent need for research on health information systems and on the inclusion of disease monitoring programs in basic health services.

3.5.4.11
Conference-Related Research

The HABITAT Agenda prepared for the second United Nations Conference on Human Settlements (held in Istanbul in 1996) provides a global framework for promoting sustainable settlements development. The Conference adopted a Global Plan of Action, which defines common goals and strategies that are then specified in national plans of action. The two major themes of the conference are shown in *Tab. 2*.

Further development of national shelter policy in individual countries should be supported through research. The Global Plan of Action refers in this context not only to the general aim of adequate shelter for all:

- It is also essential to take into consideration the needs of the poor, the homeless and ethnic minorities; measures to eradicate poverty must be integrated in human settlements and shelter development policies.
- To improve the data base, it is necessary to develop a housing information system.
- Intensified research must be carried out with the aim of improving land management and land tenure policy (safeguarding rights to land).
- Human settlements are increasingly vulnerable to natural and man-made disasters. Such crises are all the more severe in those countries which have a

Adequate shelter for all	Sustainable settlements development in an urbanizing world
National shelter policy	Sustainable land use
Organized shelter delivery systems	Combating poverty and creating jobs
Ensuring access to land	Environmentally sustainable and healthy human settlements
Basic infrastructure	Sustainable energy use
Improving construction and maintenance	Sustainable transport and communications systems
Vulnerable groups	Conservation and rehabilitation of historical and cultural heritage Improving urban economies Balanced development of rural settlements Disaster preparedness, crisis management and post-disaster rehabilitation capabilities

Table 2
Key strategies for implementing the Global Plan of Action adopted by the Habitat II Conference in 1996.
Source: UN, 1996

particularly low capacity to respond to crises. Research projects addressing the technical, social and economic aspects of rehabilitation measures and the development of effective response strategies and guidelines are absolutely essential in this connection (*see Section B 1*).

The research fields listed above must be viewed in their overall systemic context – improvements to buildings must be adapted to the physical conditions at the specific locations, for example (risk of earthquakes). Interdisciplinary research and development activities aimed at identifying and analyzing critical trends regarding settlements, settlements policy and settlements programs will thus remain an important function of HABITAT. The importance of data collection and management in the fields of population trends, migration and urbanization was stressed by the 1995 World Summit on Social Development in Copenhagen, and is sure to be dealt with at the 1996 World Food Summit in Rome. Implementing the research activities that have been called for in the context of these conferences will further our understanding of the complex interactions that global change involves, and will reveal new ways to solve the problems that humanity now faces.

3.6
Economic Research

3.6.1
Relevance of the Economy for Global Change

One of the major principles and objectives of AGENDA 21 is to integrate environmental protec-tion and economic processes in such a way that sustainable economic development is ensured. The concept of sustainable development still needs to be operationalized, a challenging task for economic research as well. Of particular relevance are the questions concerning the identification of the essential (i.e. non-substitutable) elements of "natural capital", the substantiation of demands for intergenerational equity, and the refinement of the sustainability principle to include the economic and social dimensions.

The primary concern of economic research relating to the global system was and is the explanation of key trends within the global economy, the activities of relevant actors including states and international institutions, and the search for concepts geared to ensuring the natural basis of sustainable livelihoods. Globalization of the world economy and internationalization of production are leading to economics shifting its orientation from the national to the global level. Globalization refers here not only to the geographical expansion of sales and procurement markets, but also the homogenization of consumption and production styles, the dominant role of the industrialized countries in shaping the main direction of technological development, the growing complexity of ownership patterns and the standardization of important legal frameworks.

Compared to other areas of research, especially in the natural sciences, economic research on global environmental problems is a relatively recent development. Although it has clearly increased in significance, especially through the climate protection debate, there are still no summary descriptions of this research field, with the result that the following analysis should only be seen as an attempt to derive some key approaches and neglected research areas from the range of activities currently undertaken.

There is general consensus among economists that the discipline of economics could provide some important contributions to global change research: firstly with respect to the *identification of causal factors* and the *analysis of the effects of global environmental stresses and environmental policy*, secondly, the operationalization of the sustainability principle, and finally as regards the *instruments* for influencing supraregional trends.

Another important field involves economic research on regional phenomena of major relevance to global change. Depending on the region and prevailing environmental conditions, however, this research can lead to totally divergent results: for example, an economic assessment of soil erosion in Costa Rica, which has fertile soils, would produce results totally different from those for the Sahel region (WBGU, 1993).

Two issues are particularly relevant for the promotion of German research:

1. In which economic research activities with a regional focus does Germany intend to participate in future?
2. What role is played by economics within research activities concerning the global environmental changes that affect Germany and its neighbors?

One example here is a climate impact research project financed by the BMBF that examines the effects of sea-level rise on the North German coasts (*see Box 8*).

A less specific role is played by global environmental economics where the implementation of global agreements at national level is concerned, in that such instruments differ little from national policies in essence (e.g. an EU-wide CO_2/energy tax, which would serve both national and international goals). What is relatively new and less widespread in other countries is the most recent debate on rule-oriented approaches (*Ordnungspolitik*) for managing global environmental problems.

3.6.2
Major Contributions by German Economic Research

The contribution made by the economic research community in Germany to explaining global change and analyzing the environmental implications of the trend towards economic globalization has been modest by comparison. On the whole, economics did not become involved in the analysis of global changes until fairly late on account of its traditional focus on national issues (national economics). This is particular-

ly the case where German economics is concerned, which was primarily concerned for a long time with assimilating Anglo-Saxon research results and which in a certain sense reflects trends within the latter after a time delay (the virtual absence of a "colonial tradition" may play a role in this respect).

Only recently has there been an opening of sorts, although one could not maintain that German economic research has provided any decisive impulses for the field of global economics. This is even the case for such classical topics as environment and free trade, analyzing the global origins and dissemination of technological advances, the relative constancy of large-scale disparities in wealth, or divergent forms of economic networking (triad formation, for example). The analysis of the costs involved in protecting the environment or for delaying such action has also been more national than international in focus. If there is a specific German contribution of any importance, this would be the discussion on regulatory aspects initiated by Walter Eucken and Friedrich A. von Hayek. The latter enjoyed a noteworthy revival in the context of the sustainability debate (Gerken and Renner, 1995; RWI, 1995; IAW, 1995; IÖW, 1995; ZEW 1995) and led to some interesting attempts to operationalize the sustainability principle, to a reappraisal of international organizations and to important ideas regarding the ecological renewal of the market economy. Also interesting are game-theoretical approaches for explaining the behavior of globally relevant actors. The German contributions to resource economics have been largely aligned to Anglo-Saxon research; there has been a special development in Germany in the environmental policymaking field, by contrast, mainly as a result of the debate over materials control policy. Syndrome-based research approaches may enable the German research effort to develop a greater degree of originality than has been the case so far.

3.6.3
Integration of German Economic Research in International Programs

Before assessing German involvement in international research programs, one must realize that there has been little in the way of globally oriented research activities dealing explicitly with global change and with special focus on the subject of economy and environment. The only groups to address these issues are those set up in the context of climate protection policy (IPCC), which deal with the costs incurred through changes caused by the greenhouse effect and with the cost-benefit analysis of response options for climate protection (adaptation and/or mitigation), the United Nations groups working on the operationalization of the sustainability principle and its integration into national accounts, and OECD working groups addressing issues relating to the globalization of the economy and the environment. However, it would be wrong to speak of any major influence on the part of German economists in these panels and committees up to now.

3.6.4
German Economic Research – Research Needs Concerning Global Change

3.6.4.1
Scientific Issues

ENVIRONMENTAL POLICY TARGETS

As a result of globally divergent preferences, national interests and economic and ecological problems, defining goals for protecting the global environment is bound up with major difficulties regarding economic valuation that involve, *inter alia*, the assessment of different kinds of goods and assets, and which therefore must lead to the application of comprehensive cost-benefit analyses to response options under negotiation. The major consensus among global actors regarding the importance of the sustainability principle as a basis for action must not be allowed to conceal these difficulties. Satisfactory operationalization of this latter guiding principle, in a manner that produces general agreement and which delivers accepted conclusions for concrete environmental action, has yet to materialize.

A major dilemma has to be resolved when deriving global targets, namely the divergence between individual, national and global rationality. If a person wishes to make an individual contribution towards solving global environmental problems, he or she must often circumvent several "rationality traps" at once, because at the level of individual rationality the costs associated with a more "ecological" form of production can exceed by far the benefit that the individual derives from "global environmental protection" as a public good. For individual countries, there is little rationality in waiving welfare gains in order to contribute to climate protection if the effect on climate is negligible and the costs for such a contribution are high. Within such a framework, action may be dominated by the interests of those domains and sectors that suffer negative impacts, and shape the behavior of states at international negotiations. It is all the more important, therefore, to identify a system of objectives that generates discernible benefits at different levels of action. For researchers in the field

of ecological economics, it would therefore be important to collaborate more intensively with those disciplines that address fundamental normative issues, philosophy and ethics especially, but also the natural sciences, which play an important role in the debate over environmental standards.

Ever since the UNCED Conference in Rio de Janeiro, at the latest, there has been visible acceptance of the sustainability principle. However, this raises the question of how to operationalize the concept at the policymaking level. In this connection, research should focus primarily on the following basic issues:

Issue 1:
What people-society-environment relations are of critical importance in the context of global change?

Issue 2:
What elements of "natural capital" are of essential importance (i.e. non-substitutable) at which spatial level (global, transnational, national, regional), and hence require special protection (valuation of environmental assets)?

Issue 3:
To what extent are such definitions determined by differences in preference? How do individual and politically relevant preferences arise?

Issue 4:
What weight should be attached to long-term intergenerational equity? How can distributive justice best be defined? Would this then apply to all conceivable population trends?

Issue 5:
Does the concept of sustainability require further differentiation with respect to specific problems (geographical areas, sectors, companies)? What distinctions need to be made?

Issue 6:
Is it possible to operationalize and render more specific the existing rules for the use and management of resources in order to achieve sustainability? Do all management rules obtain the same weighting?

Issue 7:
To what extent does the syndrome approach recommended in this Annual Report lead to a solution of the operationalization problem?

Issue 8:
What are the implications for rule-oriented politics of the sustainability principle and its various interpretations?

Issue 9:

Does the sustainability principle need to be enlarged by a social and economic dimension? What are the interrelationships between ecology, economics and social acceptability?

Issue 10:

How can sustainability be expressed in terms of indicators?

PEOPLE-SOCIETY-ENVIRONMENT RELATIONS

Uncertainty about important aspects of *people-society-environment interactions* (Issue 1) and about key parameters of global relevance explains some of the problems encountered in the debate over policy objectives. The economic and ecological domains are now understood to be closely linked sub-systems of a superordinate system that could be described by means of a comprehensive theorem or model. However, some of the modules needed for analyzing global environmental targets, world economic growth, population growth, the international distribution of income and for discounting the interests of future generations have not been developed. It has been virtually impossible to calculate with any precision the national, regional and intertemporal costs of protecting, or failing to protect the environment.

Another important research focus concerns the feedback effects of environmental policy decisions on the social system. Experience gained from the oil crisis of the 1970s, when oil prices rocketed, show that international shifts in purchasing power, such as those anticipated from a drastic increase in energy prices, may produce the intended ecological effects under certain circumstances, but can also have severe impacts on world trade – with developing countries being the worst affected. This makes it more difficult to formulate targets and objectives, or to disclose national preferences, and explains at least in part why the sustainability principle has not led to anything more than a general definition of important management rules (Enquete Commission, 1995b; SRU, 1994), which themselves involve a number of problems. What is needed, therefore, is more socioeconomic research with a regional and sectoral resolution. In particular, the scientific foundations for a comparison of the (regional) costs of protecting and not protecting the environment need to be improved.

THE VALUATION OF ENVIRONMENTAL ASSETS

Current methods for monetizing environmental assets (issue no. 2) are in need of refinement. Besides the well-known indirect methods such as hedonic pricing or the "travel expenses approach", attention has centered more recently on direct methods referred to in general as contingent valuation, i.e. procedures in which the valuation of environmental assets is carried out using survey techniques. Most research aimed at enhancing these valuation methods has been carried out in the USA, not least on account of their increasing practical (and hence environmental) relevance in the context of American liability law. A number of studies have meanwhile been published which attempt to define standards for valuation analyses in order to ensure maximum reliability and validity of results. The Report of the NOAA Panel deserves special mention in this connection (NOAA, 1993; Bateman and Turner, 1993; Mitchell and Carson, 1989).

German economists should involve themselves more in the development of such research approaches. On the one hand, problems concerning the application of contingent valuation demands greater research efforts regarding, for example, the differentials between the willingness to pay for benefits or to accept compensation for costs. Another issue which is gaining in importance concerns the extent to which existing valuations can be applied to similar cases, thus reducing the number of cost-intensive studies (Pruckner, 1995; OECD, 1994). Secondly, efforts should be made to apply familiar valuation methods to global environmental problems. Valuation studies at international level have related primarily to the greenhouse effect, so the Council recommends that the focus be extended to other global environmental problems, such as marine pollution, soil degradation or biodiversity (which involves special valuation problems).

The sustainability principle is based to a decisive degree on the hypothesis that so-called "natural real capital" cannot be replaced indefinitely by artificial, i.e. man-made real capital. This hypothesis has to be specified in greater detail, especially as regards the definition of what constitutes essential natural capital (UGR, 1995). Our understanding of "non-substitutability" varies, depending on the spatial context. It is therefore important to reach agreement on what comprises the global commons, because such demarcation lines are always determined in part by valuations and preferences.

THE ORIGINS OF PREFERENCE

This leads to the third issue, which concerns the origins of preferences. Monetary valuations mostly result on the basis of individual preferences and/or willingness to pay. However, individuals are often aware of their preferences for environmental assets to a limited degree only. For this reason, the way in which information is presented in surveys can influence the results of valuation. Research should therefore be conducted into how the information provided in the valuation study is processed by the individuals involved (Pruckner, 1995).

Lack of knowledge about how a specific preference for environmental assets arises makes it difficult to say whether preferences (as opposed to the perception of same) are changed or manipulated as a consequence of the monetization methods used (Weimann, personal communication). As long as the relation between the procedures for identifying preference and the formation of preference itself remains obscure, the results of monetization studies must be viewed with caution. Research must acquire an interdisciplinary dimension if progress is to be achieved in this area. The aim should be to integrate psychological and social psychological components in particular. This is all the more so where global environmental problems are concerned, because changing environmental quality is something that individuals are often unable to perceive (climate and ozone concentrations being a case in point).

It is also important to explain the *national preferences* which shape the behavior of states at international negotiations and which can lead to divergent interpretations of the sustainability principle. This aspect has major relevance, in that agreements committing sovereign states to cooperative behavior can only be achieved when participation and continued presence in the coalition is voluntary. There has been various research work on this aspect, mainly empirical studies in the field of political science, but from the perspective of economic theory, the question as to how stable coalitions of several countries can be achieved has gone largely unanswered (Barrett, 1991 and 1993; Weimann, personal communication).

INTERGENERATIONAL EQUITY

The fourth issue relates to intergenerational equity. The sustainability principle has a strong ethical dimension, as expressed in the demand for interregional and above all intergenerational equity. The demand for intergenerational equity, mostly based on Rawls's theory of justice (Rawls, 1972), requires that the range of material options open to (or level of assets enjoyed by) future generations be at least at high per capita as those of the current generation, in order to ensure that all generations have the same opportunities to develop. The literature on the subject, as already mentioned, distinguishes between stocks of natural and "artificial", i.e. man-made capital. If one assumes limited substitutability between these stocks and prioritizes the demand for intergenerational equity, the consequence is to demand the preservation of natural assets or those elements which are considered to be non-substitutable.

Such an analysis raises the question of how to weight the interests of different generations. Attempts to operationalize the concept automatically involve the question of discounting. Assuming strict complementarity between non-renewable natural capital and renewable capital, i.e. real capital produced by humans, then a social discount rate greater than zero will always imply that the needs of future generations are being assigned a lesser value than those of the current generation. Thus for many analysts, discounting is only permissible in respect of substitutable stocks and/or the yields obtained from them. If one deviates from the complementarity hypothesis and assumes that stocks are substitutable, then the controversial question is raised as to the "correct" intergenerational discount rate (Lind, 1990; Freeman, 1993). If one also addresses the question as to which needs are to be taken into account and whether this equity requirement shall apply for all population growth rates in the future (RWI, 1995), then the problems of operationalizing the sustainability principle are complicated still further. The issues referred to here require greater collaboration between the fields of economics, on the one hand, and ethics, sociology and psychology, on the other.

Another factor of critical importance when resolving the conflict of interests between the generations, besides the actual discount rate to be chosen, concerns the calculated costs and benefits of environmental degradation or protection that are imposed on or acquired by the various generations (which are then subject to discounting). In the view of the Council, research should be concentrated more heavily on determining these variables, i.e. the costs to the national economy of abatement measures in the present and the costs of environmental degradation imposed on future generations. While further research efforts on discounting are certainly necessary, the Council considers it far more important for the resolution of intergenerational conflict of interests to close the gaps in our knowledge regarding the economic impacts of global environmental change. At the same time, research should address the fundamental issues underlying the somewhat controversial equity requirement based on Rawlsian principles.

DIFFERENTIATION OF THE SUSTAINABILITY CONCEPT

The fifth issue concerns the necessity to make further *distinctions in the sustainability concept* as its relates to specific problems. The existence theorem or sustainability principle requires various kinds of differentiation depending on the environmental problem at issue. In its work to date, the WBGU has filtered out the following problems for separate analysis: global climate changes, depletion of the ozone layer, soil degradation, biodiversity loss, overexploitation and pollution of the oceans, and water scarcity and pollution. Our studies have shown that each complex needs its own specific regionalization. Whereas the climate problem allows a global definition of sustainability criteria with respect to the maximum permissible emissions of greenhouse gases, coping with soil degradation or drinking water scarcity requires the derivation of regional indicators. Thus there is still a substantial need for research in this field. Disputes also surround the question whether sectoral or even microeconomic differentiation is needed alongside regionalized analysis.

UTILIZATION AND MANAGEMENT RULES

The sixth issue concentrates on deriving *rules for resource management*. Of special interest in this respect is the refinement of criteria for the utilization of exhaustible and renewable resources. An integral part of the sustainability concept is the rule that finite resources may only be consumed to the extent that such consumption is offset by the creation of substitutive resources through research and development, but this rule involves special problems and is virtually impossible to realize in practice (Cansier, personal communication; Klemmer, 1996). Moreover, the innovative capacity of industry and society to develop substitutions for scarce resources tends to be underestimated.

Similarly, the basic management rule which states that consumption of renewable resources be limited to their regeneration rate is difficult to implement. This would involve valuation of reserves, and would be virtually impossible to achieve within an internationalized division of labor. The same problem arises when trying to define the global commons and its components. Whether realistic utilization rules can be formulated for the conservation of a multifunctional resource and what shape these rules would actually take is still an unresolved issue (Cansier, personal communication). Finally, critical thought must be devoted to the introduction of risk factors and assessments.

OPERATIONALIZATION

The discussion so far on the goal of sustainable development shows that this ubiquitously accepted imperative is subject to a wide range of interpretations, which make is very difficult to achieve operationalization at the global level. The question whether agreement could at least be reached on certain forms of non-sustainable trends is thus all the more interesting. The Council's own syndrome-based approach points in this direction and may make it easier to resolve the problem of operationalization (*Section*

C 2). To that extent, research should concern itself with this approach more in future (issue no. 7).

IMPLICATIONS FOR RULE-ORIENTED POLICY

The *regulatory implications* for rule-oriented policy of the sustainability principle have been largely neglected until now (issue no. 8). The questions as to whether the sustainability principle and its various interpretations are reconcilable with the market economy, whether the market economy can even be brought to a stronger long-term orientation or explicit concern for ecological issues, and with which instruments ecological goals can be achieved while still conforming to market principles, have all been treated as marginal concerns until now (Rentz, 1994; Brenck, 1992). The recent past has seen a growing number of studies focusing on these issues (Gerken and Renner, 1995; RWI, 1995; IAW, 1995, IÖW, 1995, ZEW, 1995). A striking feature of these studies is that the demand for an efficiency and sufficiency revolution based on ecological and equity considerations, as put forward in some interpretations of the sustainability principle, assigns major responsibility for controlling these processes to the state. If this "revolution" is accomplished, for example through constant increases in the taxation of energy and other resources (Görres et al., 1994; DIW, 1994), then market prices will be increasingly determined by the state, which may thus involve a gradual transformation of the entire economic system.

There is an urgent need for further research to determine the extent to which existing concepts for an eco-social market economy comply with the sustainability principle and yet remain in conformity with the system. Recent research results from the field of "evolutionary environmental economics" or "new political economics" must be taken into account here. The failures of markets and policies must be weighed up against each other. The issue in all cases is how to achieve a long-term orientation at the microeconomic and political decision-making levels within the market economy and the democratic system respectively. A desirable path would be to succeed in motivating consumers to base their actions on environmental considerations and long-term perspectives, and to set in train, through their willingness to pay, an "ecologization" of the economy. Related issues concern the relationship between free trade and environmental protection, or the demand for maintenance or enhancement of national competitiveness (*see Section B 3.6.4.2*).

SOCIAL AND ECONOMIC DIMENSIONS

Closely connected with these fundamental regulatory issues and the indicator problem dealt with below is the question of the extent to which the sustainability principle must embrace a *social and economic dimension* as well (issue no. 9). In the more recent debate in Germany, especially within the German Parliament's Enquete Commission Protection of People and the Environment (Enquete Commission, 1995b) and the regulatory policy context (Gerken and Renner, 1995; RWI, 1995; IAW, 1995, IÖW, 1995; ZEW, 1995), there is a discernible tendency towards such an extended sustainability principle. Mention is often made of the Three Pillars Model, for example. The problem of how to operationalize the principle of sustainability is thus acquiring additional contours, in that concepts such as economic compatibility admit a similarly wide range of interpretation (Klemmer, 1994). Moreover, the issue is raised as to the inter-relationship between environmental soundness, social acceptability and economical compatibility. In its statement to the Berlin Climate Summit, the Council proposed a method for integrating the ecological and economic dimensions of sustainability that also reveals the major gaps that still exist in economic and social science research (WBGU, 1995b).

INDICATORS

Research on indicators (issue no. 10) is still in its infancy, as shown by the numerous attempts to elaborate practicable indicator systems in the context of AGENDA 21 and the work of the CSD (Commission on Sustainable Development). These attempts have brought to light a large number of conceptual gaps, thus underscoring the substantial need for research in this field. More should be known about

- different options for defining specific quantitative indicators for the utilization of environmental assets and resources. An important aspect here concerns the question of how far "objective" minimum standards can be derived as "conditions" for sustainability. In this case, sustainability could be defined in terms of a corridor between certain minimum standards or "crash barriers" (*see Section C 2.1.2*).
- methods for determining avoidance costs and for developing practicable and adequate concepts of ecological national product.
- the applicability and strength of long-term deviation indicators, whereby the sustainability principle requires long-term conservation goals as normative reference points.
- the development of non-monetizing aggregation methods (especially the suitability of the rates-to-goals approach for deviation indicators).

Special attention must be devoted in any event to accumulative stresses on the environment and the irreversibility of some factors.

BARGAINING PROCESSES

Successful management of global environmental problems depends to a critical extent on the specific conditions under which states are willing and able to enter stable coalitions in which they commit themselves to implement measures that do not necessarily produce direct benefits for the individual country, but which are desirable for the coalition as a whole. This is an important question, insofar as the actors at international level are predominantly sovereign states, the principle of voluntary coalition prevails and the stability of coalitions cannot be forced at all or only rarely on states.

So far, economics has produced only rudimentary frameworks for analyzing the formation of stable coalitions, the successful operation of conventions, the assignment of national sovereignty to institutions, etc., whereas political science has already made some impressive advances in this area. If economic analysis is to achieve any progress here, then game-theoretical approaches for describing negotiation equilibria and economic options must be further refined. The solution to many global environmental problems often involves modifying the conditional framework in such a way that equilibrium and optimization are congruent. The transaction costs of international agreements must also be examined, since they vary according to the specific environmental problem in question.

It has been pointed out in this connection (Endres, personal communication) that the explanations of reality put forward by economists do not always coincide with those of legal experts. Greater cooperation is therefore required between economists and the latter, in addition to the cooperation with political scientists already referred to above. Answers are needed to the following questions in particular:

- How significant is the much-cited "free-rider" problem in reality?
- How useful are the response hypotheses used by economists when applied to certain forms of behavior during international negotiations?
- Are hypotheses about the connection between lack of compliance and failure to conclude agreements (anticipation hypothesis) tenable?
- Is it possible to formulate economic explanations for the divergent preferences of states for certain environmental policy instruments?
- How significant are compensation payments? Can the core issues of environmental negotiations be linked to others ("package solutions") in order to reach a successful outcome?
- How must sanctions against violations be designed?
- To what extent does global change require centralization (e.g. an Environmental Security Coun-

cil) in order to achieve results? Is the model of competing regulatory frameworks a real alternative? Does the economics of federalism provide a basis on which to base solutions to the agencies problem?

- How should institutions like the World Bank, the UN or the WTO be assessed ecologically? What ideas for reform are supplied by the social theory of bureaucracy?
- Is a Second Chamber or Federal Bank of the Environment needed at national and/or international level in order to enforce a long-term orientation?

INSTRUMENTS

There continue to be shortcomings, for a variety of reasons, in the analysis of *global environmental policy instruments* (*see Section C 7.5*). Firstly, policy instruments have been neglected in academic debate; secondly, there are no global models for simulating the feedbacks of environmental policy decisions on key global parameters such as energy prices, exchange rates, world trade, etc. Given the absence of any superordinate "world government" with the power to deploy its own instruments, agreement on policy instruments must be engendered through negotiations between independent states, and an incentives system to promote this process must be created. At international level, this inevitably means some kind of compensation or transfers. Both measures produce certain economic and ecological effects.

The debate on policy instruments is currently focusing on three different approaches: tradeable quotas and permits, joint implementation, and taxes and levies. The former two areas have been the subject of recent feasibility studies (Heister et al., 1990; Huckestein, 1993; Maier-Rigaud, 1994; Endres and Schwarze, 1994; Hansjürgens and Fromm, 1994; Fromm and Hansjürgens, 1994; UNCTAD, 1995), while others are in progress. Interest centers above all on the political effectiveness of the relevant instrument and compatibility with existing law. Market simulation and the analysis of efficiency and incidence are dogged by methodological problems. The WBGU has stated its position on joint implementation on several occasions, but even though other analyses have been produced in the meantime (Michaelowa, 1995; Rentz, 1995; Heister and Stähler, 1995), there are still some areas where research is needed.

The introduction of a national energy tax has been the subject of several studies (Conrad and Wang, 1993; DIW, 1994; Enquete Commission, 1990 and 1995b; EWI, 1995; Karadeloglu, 1992; Oliveira-Martins et al., 1992; Standaert, 1992; Welsch and Hoster, 1995; RWI, 1996). The problem with the latter is that

they mostly define the conditional framework at global level in a "manual" fashion (energy price trends, for example, or exchange rates and volume of world trade), are confined to comparative, statistical methods, neglect distributional effects and contain imprecise assessments of structural impacts. The time needed to implement such measures and the adaptive processes involved are inadequately described, as are the concomitant effects on employment. The positive effects, such as higher employment and the reduction of emissions, are most likely underestimated. Modeling intertemporal equilibrium could lead to methodological improvements in this respect. What is especially important is that global analyses take account of feedbacks on global energy prices (Ströbele, personal communication). All in all, thinking on environmental policy instruments has been based until now on rather shaky foundations.

When analyzing the agencies and instruments of global environmental policy, the interest of environmental economists in the field of global change research has centered primarily on the climate change problem and, to a much lesser degree, the loss of biodiversity. The Council sees an urgent need to extend environmental economic research to other global environmental problems, such as protection of global water reserves and soils. Research in these areas exhibits major deficits.

3.6.4.2
Economic Research in Specific Policy Fields

OVERVIEW

The research assessment in this section has mainly focused on the economics of climate change, the world's oceans and soil degradation, which are all constituent elements of global change research. But there are also numerous economic analyses covering specific fields of national policy that, while distinct from environmental policy, are nevertheless intimately related to it. Transport policy, energy policy, agricultural and forestry policy are prime examples. These policy fields have their own impacts on the global environment, and – conversely – are themselves the subject of global environmental policy, since the issues involve national and local implementation. Examples include the generation of CO_2 emissions in the energy and transport sectors, or the potential of forestry measures for enhancing sinks.

Coordination problems therefore exist between global and national environmental policy, on the one hand, and policy-based research or advisory activities, on the other. A policy for energy saving, to take perhaps the most important area, is aimed at attaining both regionally and globally defined goals for environmental quality, especially when coordinated at international level. Use of such instruments varies comparatively little according to global or regional situation – greater reductions of nitrogen oxides at the regional level and greater CO_2 reductions at the global, for example – so it cannot be the responsibility of an advisory body for global environmental change to formulate the need for research on national energy policy; reference is made instead to the relevant work of the Council of Environmental Experts (SRU, 1996).

One topic in particular should, however, be addressed. In its 1995 Annual Report, the Council focused in detail on the interrelations between global environmental problems and the international trade regime. The two years that have elapsed since completion of the Uruguay Round have witnessed a number of changes in this respect – work is now being carried out by the newly-established Committee on Trade and Environment, for example. This prompts us to focus once again on the need for research in this area.

THE NEED FOR RESEARCH ON TRADE-ENVIRONMENT PROBLEMS

International networking of national economies, and hence the volume of world trade in goods and services, is constantly on the increase, with manifold consequences for the natural environment at global and regional level (*see* WBGU, 1996 for more detail). This particular complex involves many different aspects and issues ranging from regulation of competition (eco-dumping), internal organization (dispute settlement procedures), institutional frameworks (creation of a new environmental organization and inclusion of environmental aspects in the work of the WTO) and political questions (legitimation of non-governmental organizations within WTO negotiations).

The number of scientific publications on the subject of "trade and environment" may have grown substantially since the early 1990s, but these are largely confined to theoretically abstract analyses, and originate almost exclusively in Anglo-Saxon countries. In the view of the Council, Germany's leading position in the world – in terms of both foreign trade volume and the progressiveness of its environmental policies – requires that German research devote much greater attention in future to the interrelationships and interactions between free trade and environmental protection.

Public interest in these questions was considerable in the early 1990s in the context of the NAFTA negotiations and the Uruguay Round, but has markedly declined in the intervening period. The Council considers it essential to concentrate the trade and envi-

ronment debate on those economic and environmental policy problems that are currently most pressing. This debate should be furthered by emphasizing the selected issues described below.

SCIENTIFIC ISSUES

There is every reason to believe that the more effectively the Member States internalize externalities, the less the GATT/WTO regime will be ridden with trade-environment conflicts. If globally relevant externalities were fully internalized, the ecologically positive effects of free trade would come to the fore. The world economy has a long way to go before any such goal could be reached, and this implies a multitude of research tasks in the future. For example, studies are needed to determine what the potential responsibilities and functions of the WTO might be in connection with internalization through the implementation and/or harmonization of economic instruments.

There is also a need for much more precise information regarding the *effects of environmental policies on competition*. Existing studies covering specific sectors or industries and the effects of environmental levies on international competitiveness must be enhanced and improved. Precise analyses are needed to determine the specific cases in which environmental measures impair the competitiveness of certain industries so much that the macroeconomic consequences for income and employment are no longer acceptable. Under what conditions are the foreign trade effects predominantly positive, e.g. through innovations and investments in environmental protection? The extent to which existing levy systems, e.g. those in the Scandinavian countries, have had impacts on the competitive environment of individual countries must also be subjected to analysis.

The same applies, conversely, for the *environmental impacts of trade measures*. A list of criteria needs to be drawn up for assessing the spatial structure of the global environmental impacts induced by increasing trade. Studies are also needed on individual sectors and industries. No studies have been conducted yet on the environmental impacts of export industries under different environmental policy scenarios, to take one example.

Given the special difficulties in achieving international agreement, and the financial constraints imposed on many states and political institutions, it is particularly important to formulate and conduct research into "win-win" policy fields, i.e. it is necessary to identify those areas where free trade and environmentalism are mutually supportive. The reduction of subsidies that have detrimental effects on the environment would be one example. Many forms of resource use are heavily subsidized. Key tasks for future research are to analyze the effects of subsidies on environmental quality and free trade, and to develop policy-based approaches for dismantling subsidies.

In the latter context, researchers would have to examine the potential role to be played by the WTO as regards the dismantling of environmentally unsound subsidies. The Uruguay Round may have been an important first step towards the elimination of problematic agricultural subsidies, but the latter continue to cause major distortions of trade and the detrimental effects on the environment they induce (the Common Agricultural Policy in the European Union being one example). A key issue here, and one that needs researching, concerns the interrelationships between the world food trade and food security.

Another area in which environmentalist and free trade interests may be congruent involves the *removal of market access restrictions*, an issue to which the OECD will have to give a great deal of thought in future. One example involves so-called "customs escalation", i.e. the fact that products are subjected to higher import duties the higher the degree of refinement. Such practices probably have negative environmental impacts, because the countries affected are forced to exploit their raw materials more intensively as a result. What is needed here is a precise analysis of the raw materials concerned: the anticipated effects must be quantified, the environmental impacts of processing in the countries of origin compared with those in importing countries and the transport intensity of raw materials compared with that for processed products.

Another very topical issue with political relevance concerns the relationship between the WTO regime and international environmental conventions. The Council addressed this issue in its 1995 Annual Report.

STRUCTURAL IMPROVEMENTS NEEDED

As already mentioned, German research in the field of environmental economics has mainly been conducted in small, decentralized facilities. The international and interdisciplinary dimensions of the "trade/environment" complex means that environmental economic research will not only have to widen its scope, but must also be reorganized accordingly.

The German research community should be represented in international programs much more than has been the case, and should play a key role, for example by initiating the relevant priority research programs within IHDP. Cooperation within the scientific community should be given greater support. In this connection, the Council recommends that international workshops on the subject of Trade and Environ-

ronment, analogous to the climate initiative of the German-American Academic Council, be initiated and actively shaped.

In tertiary education, a graduate college embracing this field could prepare young researchers for the challenges that will be faced in the future. The relevance of research for practical and political action is of central importance here, so efforts should be made to establish close links with international institutions (WTO, UNCTAD, etc.) and industry through in-service training.

Other questions relate to how globally oriented economic research can become more institutionalized, and whether an interdisciplinary subject called "Global Change" should be set up at the universities. The construction of global economic models could be facilitated if a major institute were to become involved in this area. In view of the variety of methodologies and hypotheses, and the normative character of many theories, parallel research in economics would be a plausible option. In any case, there is an urgent need to promote collaboration between existing institutes (e.g. by setting up the relevant sections at "Blue List" Institutes). As far as tertiary education is concerned, "Global Change" is more suitable as a subject for graduate colleges or advanced degree courses than for undergraduate courses, where the prerequisites have first to be met within the disciplines concerned.

3.7
Research on Societal Organization

The following section examines the institutional and legal aspects of global change as they are studied by political science and law, leaving the "Psychosocial Sphere" to *Section B 3.8*, where the focus shifts to research on environmentally relevant behavior and alterations of behavior.

3.7.1
Relevance of Political Science and Law for Global Change

Environmental research within the discipline of political science, which analyses activities, processes and institutions of environmental policy, is increasingly recognized as an important form of environmental research. This also holds for global environmental politics, where strategic action on the part of political actors is particularly evident. Greater demand for political analysis runs parallel to the growth of environmental research within political science, especially in the field of international relations. However, political science research on global environmental change has only developed to any significant extent in Germany in respect of a small number of publicly debated issues, first and foremost the problems of ozone and climate change; systematic research in other areas, such as global biodiversity, soil and water problems, is still in its infancy.

Global environmental problems also account for a growing proportion of the literature on international law. Attention is centered, as in political science, on international institutions for protection of the environment, i.e. the legal relations between the states involved. There are now around 800 intergovernmental agreements relating to the environment, ranging from bilateral agreements to regulatory frameworks with universal validity. The main thrust of international law and political science research has been the study of regulatory mechanisms for the better-known environmental problems, particularly the conventions for protection of the climate and the ozone layer. The use and protection of the hydrosphere is another traditional topic within the literature on international law, because the inherently "transnational" character of the global hydrological cycle led at an early stage to the establishment of an international law of the sea and an international legal framework governing transboundary waters.

3.7.2
Major Contributions by German Political Scientists and International Lawyers

3.7.2.1
International Regimes as a Field of Research

German researchers in the field of international law have meanwhile produced a substantial body of work on legal aspects of global change, ranging from the analysis of specific conventions and treaties to the definition of general rules for international environmental law. One important publication in this context is the Yearbook on International Environmental Law, a valuable supplement to the more traditional journals in international law. In the view of the Council, however, there is still a considerable need for research, particularly as regards fundamental reforms of international law. This is dealt with in greater detail below.

A relatively old concept within international law is that of international "regimes", which originally referred to the status of a certain area, but which is now also applied to the specific set of norms, principles and rules pertaining to a specific environmental problem. The term has been used in political science

debates since the mid-1970s, with *regime analysis* being the approach most widely used in the study of global environmental problems. Initially developed in the Anglo-American world, the approach spread to German-speaking countries above all through its application to security regimes, and since the late 1980s has been commonly used, also in Germany, in the analysis of international environmental regimes.

This political science approach focused initially on explaining the incidence and analyzing the effectiveness of specific environmental regimes. For some years now, there has been increased interest in the further development, impacts and the institutional design of regimes. Regime analysts have also shown growing interest in the interactions between environmental regimes and other institutional forms of international environmental policy (such as international organizations, environmental associations, transnational economic organizations). Research of this kind is thus moving towards broader-based analyses of the conditions and the institutional design of global environmental policymaking. One element that has been missing so far, however, is the analysis of the distributional effects of international environmental regimes, particularly in the North-South context.

3.7.2.2
Regional Foci of Research to Date

Until very recently, research in political science and international law on global environmental policies has displayed a serious regional bias towards the industrialized world, and OECD nations in particular. Research on developing and newly-industrializing countries has been confined either to comparative summaries of environmental policy in different countries or on the detailed analysis of individual countries, usually conducted by country experts with a special interest in development policy. Environmental and developmental aspects remain weakly integrated in empirical research, although UNCED has provided an important impulse in this area. Consideration must be given to the fact that governments in developing countries do not view global change purely in terms of environmental aspects, but as indissolubly linked to their own economic and social development. The result is that research activities confined to international environmental cooperation will lack comprehensiveness.

Developing countries are now receiving capacity building assistance from UNEP, as well as from Germany, in designing and improving their national environmental legislation; intensified efforts on the part of German institutions would be welcome here, given the fact that many of the recently adopted conventions must now be implemented in the developing countries, which in turn demands the building of new and more comprehensive capacities.

3.7.2.3
Approaches for the Analysis of Environmental Policy

If political science is to advance its understanding of environmental policymaking, it will have to make greater use of, *inter alia*, capacity analysis. The latter views global environmental policy primarily in terms of the social, sociocultural and institutional conditions in which environmental problems can be perceived and managed. Distinguishing between divergent potentials for environmental policymaking is essentially a projection of this basic notion, and enables the conditional framework needed for successful environmental policymaking to be developed. According to this approach, environmentally appropriate behavior may develop in highly developed countries on the basis of post-materialist values (so-called *ecology of prosperity*), but can also arise in developing countries if adequate sociocultural capacities, thriftiness or strict efficiency in the face of competitive pressures so permit (so-called *ecology of scarcity*).

In view of the economic crises in the industrialized nations and the growing contribution of newly-industrializing countries to global environmental stresses, the question is raised as to whether traditional thinking in the analysis of environmental policymaking must be revised. Rather than stressing strong economic growth as the precondition for effective action to protect the environment, greater consideration would have to be given to the importance of the real scope for environmental policymaking generated by indigenous sociocultural factors.

3.7.2.4
Sustainable Development and the Common Concern of Humankind

There is a correlation between the raising of such issues and the intensified debate on sustainable development among experts in recent years. Despite the contradictions between some sustainable development criteria (SRU, 1994), the latter reveal a shift of attention from efficiency aspects to a greater focus on sociocultural criteria of acceptability and equity. Such criteria of a sustainable, environmentally sound development of economy and society are non-contentious among environmental experts; elsewhere, however, economic debates often fail to take account of environmental aspects. The current debate in Ger-

many on the country's declining attractiveness as a location for investments and production is a case in point. The question as to how this discrepancy between "expert knowledge" and "objective politics" is to be dealt with analytically, and whether the two sides can be brought closer together, remains unresolved.

Many conceptual platforms have been developed in the field of international law for the legal assessment of transboundary and global threats to the environment, such as the postulated "human right to a healthy environment", the "rights of future generations", the duty of governments to ensure sustainable development, or the legal concepts of "global commons" and the "common heritage of mankind" (Brown-Weiss, 1992). The "common concern of humankind" concept has meanwhile received the greatest support: in 1988, the UN General Assembly declared the climate to be a common concern of humankind, a principle that was then included in the 1992 Framework Convention on Climate Change and the Biodiversity Convention and (implicitly) transferred to the older conventions on the protection of the ozone layer. The novelty of the concept means that little research work has been carried out into its specific implications, so a considerable need for research exists in this area.

In terms of organization, German research in political science and law is coordinated in a number of loose research groups comprising researchers with similar conceptual and theoretical approaches. For this reason, involvement in interdisciplinary research projects on the part of social science researchers is confined to isolated cases. One exception is the DFG priority program on Human Dimensions of Global Environmental Change. In addition to various projects centered on psychological, sociological, economic, ethnological and geographical aspects, there are three projects with a political focus, namely "Implementation of Climate Protection Policies", "Environmental Policy and Land-Use Systems" and "Negotiation and Mediation Processes in Environmental Decision-Making".

3.7.3
Integration of German Political Scientists and Law Researchers in International Programs

One example for involvement in international research programs has been the special attention devoted by the Jülich Research Center in recent years to the problem of how to verify the effectiveness of climate policies. The Center has links with the International Institute for Applied Systems Analysis (IIASA) in Laxenburg (Austria), where research has been conducted since 1993, also with German participation, on the implementation of international conventions to combat global environmental change (Andresen et al., 1995). The latter study is part of a large-scale project on international environmental regimes, and had funding until October 1996. It operates as a kind of clearing house for regime theorists working on global change. The Council considers it essential that the IIASA's expertise in the field of environmental policy research be maintained and enhanced.

3.7.4
German Research in the Field of Political Science – Research Needs Concerning Global Change

3.7.4.1
Research on Specific Environmental Problems

Climate change continues to be the most important object of global change research in the fields of political science and law. In view of the tendency of many actors to focus almost exclusively on climate issues (including ozone), it is essential to stress the urgent need to broaden the research agenda to other environmental problems. As the Council has shown in its previous reports, there are many global environmental problems that require an immediate response, but where institutional research capacities for political and legal analysis are virtually non-existent. Research is needed, for example, to determine how negotiations for the International Convention for the Protection of the Marine Environment or the Forest Protocol recommended by the Council in its 1995 Annual Report could be integrated into the policymaking process in the future, or the extent to which alternative arrangements could have the same effect in the field of global environmental policymaking.

Political science and law should turn their attention to intensified study of hitherto neglected fields, especially the core problems of global change (*Box 15*) and all issues relating to environment and development beyond the purely local context. Cross-cutting problems relating to environmentally sound behavior, i. e. the environmental aspects of everyday behavior and ongoing political action, are gaining rapidly in importance. The relationship between expanding world trade and environmental degradation is particularly crucial given the progressive globalization of the world economy (*see Section B 3.6.4.2*).

3.7.4.2
The Analysis of Political Processes

Although the fundamental necessity of environmental protection and wise management of the environment is accepted by the majority of the population in many countries, the policies that are actually implemented still tend to operate against environmental imperatives. Hence, political will-formation and the implementation of environmental policy and law remain processes that have to be subjected to systematic analysis. Rather than compiling long lists of possible measures, it is essential to analyze carefully the factors that restrict action under existing societal conditions. *Comparative process analyses* should concentrate on those cases in which similar results are achieved under different circumstances, or differing results under identical conditions.

3.7.4.3
The Study of Institutions

The Council believes that special attention should be dedicated to the study of institutions. The analysis of the effects of international regimes has left many questions unanswered. The Council's recommendations for research are as follows:
- Cross-cultural analyses of factors contributing to the success of institutional frameworks.
- Analysis of the relationship between different institutional types at different levels (international, supranational, national, subnational).
- Implementation of international environmental treaties (conventions research).
- Interdisciplinary research to identify options for international management of the global commons.
- Recommendations for the environmental reform of institutions relevant to global change (e.g. WTO, international development banks, etc.).
- Research on compliance: development of instruments for enforcing international conventions.

3.7.4.4
Communications Research

Political science research into the making and implementation of global environmental policy should focus more intensively on forms, conditions and effects of international communication (*see also Section B 3.8*). This type of communications research relates, firstly, to interpersonal communication in the political decision-making and implementation process, secondly to interactions between the mass media and political decision-making, and, finally, to the impacts of new communications technologies on the formulation and implementation of policy.

First political science approaches in the field of communications research focus on the operating mechanisms of international negotiation systems. This research is important insofar as political activity has shifted more and more from parliamentary or hierarchical institutions to pluralistic, corporatist or intergovernmental negotiation systems, in which projects can only be put into practice with the consent of all participants. This is particularly the case where negotiations aimed at solving global environmental problems are concerned. These processes have been studied mainly with regard to participation (involvement of those personally affected in the decision-making process) and mediation (negotiations supported by mediating agencies).

Against the background of globally relevant "decision making under conditions of uncertainty", the distinction between arguing (rational, knowledge-based argumentation) and bargaining (negotiating on the basis of interests) is increasingly important. Bargaining plays a special role in environmental policymaking processes because the decisions are often made in the absence of certain knowledge. Research in political science should therefore be intensified, particularly in light of its special importance for the political decision-making process, and supplemented by cooperation with psychology, sociology, economics and ethics (*see Section B 3.8.4*).

3.7.4.5
Peace and Conflict Resolution Research

Global and regional environmental problems are not only the consequence, but also the cause of inter- and intranational conflicts. Examples include the conflicts generated by the export of hazardous goods, water-related conflicts, or threats to international security as a result of environmental refugees. This growing potential for conflict is not reflected in the priority research issues of German peace and conflict research. Taking the UN Secretary General's Agenda for Peace as a reference document, the following types of project seem particularly worthy of support:

- Identification of regions with environmental conflict potential.
- Formulation of concepts for the prevention of environmental conflicts.
- Research on conflict-resolution instruments (e.g. concepts such as Green Helmets or Environmental Security Council).
- Development of response strategies for the environmental refugee problem.

3.7.5
German Legal Science – Research Needs Concerning Global Change

3.7.5.1
Customary International Environmental Law

International conventions addressing global change problems, and the general norms of international environmental law have been investigated in numerous research papers in the field of law. However, there is considerable uncertainty regarding the body of extra-treaty standards relating to *specific* global threats such as stratospheric ozone depletion. While multilateral agreements are not applicable *per se* to non-signatory states, they may, under certain conditions, acquire the status of customary international law binding on all states. However, it is not clear whether this qualitative change has already occurred in the case of key conventions such as the Framework Convention on Climate Change, the Montreal Protocol, or the various amendments and adjustments made to the latter. The Council refers in particular to two criteria defining customary international law that must be subjected to critical examination in the light of global ecological interdependencies.

Firstly, prevailing legal opinion requires a certain *duration* as state practice before a norm or standard can be defined as customary international law. This criterion may only be applicable to a limited extent, given the substantial degree of scientific uncertainty regarding global environmental changes and the sheer scale of potential damage. The Council refers here in particular to the problem of stratospheric ozone depletion, which necessitated immediate global counteractive measures as soon as the scientific community had drawn attention to the problem. Where such threats exist, the international community should consider establishing a legal framework more or less spontaneously in the form of multilateral treaties, provided that these are ratified by a very high and representative number of states.

Secondly, prevailing opinion permits a state that persistently objects to the establishment of a norm in international customary law to be released from compliance. This principle seems untenable in the face of the threats to essential ecological systems and the endangering of the stratospheric ozone layer and the Earth's climate. The definition of climate, biodiversity and (implicitly) the ozone layer as "common concerns of humankind" by the UN General Assembly and the 1992 Rio Conventions (*see Section B 3.7.2.4*) are indications, in the view of the Council, that the international community accepts a certain limitation of sovereign rights on the part of individual states where humankind as a whole is threatened – at least where the above problems are involved. In this connection, analyses must be conducted to determine whether prohibiting agents which damage the global environment – such as CFC emissions – does not already constitute peremptory norms of international law (*jus cogens*).

3.7.5.2
"Environmental Solidarity" as a Legal Principle

In its 1993, 1994 and 1995 Reports, the Council repeatedly pointed out that timely and effective measures to combat global environmental threats are only possible when combined with international support programs. Many new conventions now determine, often in literal agreement, that the industrialized countries should bear "all agreed incremental costs" for implementation of the conventions in the developing countries and that they must facilitate the transfer of environmentally sound technologies (see, for example, the Framework Convention on Climate Change and the Montreal Protocol). The issue this raises is whether these often uniform provisions, combined with the factual indispensability of support programs in times of rapid global environmental change, do not already substantiate the status of general international legal obligations. The Council believes that there are important arguments in favor of a *general duty of environmental solidarity on the part of the industrialized nations* vis-à-vis the developing countries, and that discussions should be held on the status and scope of this duty in interdisciplinary dialog between political scientists, law experts, economists and ethical philosophers (*see Section B 3.8.4.1* on distributional equity).

3.7.5.3
The Role of "Civil Society" in International Law

There have been many references in recent social science research papers to the special importance of *non-governmental actors* in connection with international environmental protection. The United Nations Conference on Environment and Development in 1992 granted a wide range of participatory rights to such bodies. The Council believes that the international community should consider assigning non-governmental actors greater legal powers in international environmental law. This would not be entirely new, as shown by the traditional rights enjoyed by the Holy See, the Order of the Knights of Malta, or the

International Red Cross Committee, and the voting rights of worker and employer organizations in the ILO. A first step would be to give NGOs the right to bring an action in the relevant dispute settlement procedures for international environmental conventions, such that non-governmental organizations, as "trustees of the environment", could sue signatories for non-compliance. This proposal is reminiscent of recent debates over the intrinsic rights of nature, which could only be realized in the form of loci standi exercised by NGOs as trustees of the environment.

Little has been done to clarify the specific design that such procedural innovations might have. However, because they can only be created through international treaties, there is considerable need for research on specific compromises that could be acceptable to the majority of states, as well as appropriate legitimization procedures regarding the participatory rights of individual NGOs. A conceivable alternative would be to create a United Nations High Commissioner for Environmental Protection and Sustainable Development, which was recommended as long ago as 1987 by the legal experts of the World Commission for Environment and Development. Such a trustee of the environment and future generations, who could be granted the right to take legal action against individual states within the framework of treaty-based dispute settlement regulations, could act in a manner similar to Scandinavian ombudsmen, passing on complaints from non-governmental organizations to the various convention bodies and functioning as a link between states and the global environment associations. This is another area where international law needs to conduct research.

3.7.5.4
Legal Issues Regarding the Impacts of Climate Change

The Council works on the basic premise that major changes to the global environment can be prevented if appropriate measures are taken. However, if preventive measures are inadequate, large-scale *adaptive measures* may prove necessary, and certain regions may suffer severe damage. This aspect must be treated by international law research as well.

There is no excluding, for example, that numerous people will have to abandon their homes due to changing climatic conditions or rising sea level. These *environmental refugees* are not yet covered by the Geneva Convention relating to the Status of Refugees. The threat of climate change is induced, however, by the entire international community – above all the industrialized nations – so environmental refugees must also enjoy the protection of the entire international community. It is therefore essential that appropriate institutional foundations for the legal status of environmental refugees be formulated before it is too late.

Another aspect concerns the enormous *cost of adaptive measures* necessary in some regions, especially against the consequential effects of climate change. There is a need here, related to the *duty of environmental solidarity* mentioned above, to clarify the responsibilities of the international community, especially on the part of the industrialized countries.

3.7.5.5
Basic Legal Principles Related to Trade Restrictions in International Environmental Policies

There is growing acceptance of trade measures aimed at enforcing environmental standards in some states and areas of activity. Examples include the extra-territorial application of the Marine Mammals Protection Act of the United States, or the control of trade with non-parties pursuant to Article 4 of the Montreal Protocol. Despite the volume of research papers already available on this issue in the field of environmental law, the Council is of the opinion that the demarcation between permissible and impermissible trade measures is not yet sufficiently precise, especially in view of the GATT/WTO regime and the possible need for reforms in this area (*see Section B 3.6.4.2*).

3.7.5.6
Institutional Foundations for Innovative Forms of Global Environmental Policy

The various innovative ways to overcome global environmental problems discussed in the 1995 Annual Report (WBGU, 1996) continue to raise numerous questions as far as specific legal design is concerned. For example, it is not clear how best to shape the international laws governing the international trade in emission permits recommended by the Council. However, some experience has now been gained with regulatory mechanisms within national law (in the state of California, for example) that could function as models in this connection (Hansjürgens and Fromm, 1994). Although the Council assumes that national experience can indeed be transferred to the international level, there is still considerable need for research concerning specific details. Hence, progress at international negotiations will probably depend to a critical degree on the extent to which international lawyers are able to develop consensual models for

policy innovations in the climate protection field, one example being an international trade in emission permits.

3.7.5.7
Further Development of Decision-Making and Dispute-Settlement Procedures

The pressing problems of global environmental change necessitate innovations in the decision-making procedures, implementation mechanisms and dispute settlement procedures of international conventions. These innovations should take the form of limited and precisely defined restrictions on national sovereignty *within* convention regimes, as is already the case with the qualified majority decisions permitted by Article 2 of the Montreal Protocol, for example. The Council appeals for wider application of the majority principle in environmental conventions, embracing the group veto rights of North and South, the participation rights of non-governmental organizations and the binding jurisdiction of international courts or courts set up to regulate specific conventions – a Climate Court, for example.

3.8
Research on the Psychosocial Sphere

3.8.1
Relevance of the Human Sciences for Global Change

Anthropogenic global change results, as the term suggests, from interactions between people and their environment. Human activities, however, are not only the cause of the changes occurring, but are also affected by such changes at all times. Moreover, humans can respond to or take precautions against global change in order to influence the changes themselves or to adapt society to them. To that extent, global change is a *behavioral problem*, and environmental crisis is a *crisis of society*. If the probability is high that global climate change, for example, is dependent on human behavior, and if it is also known which greenhouse gases are primarily responsible, then an important research objective – also for environmental policy – is to identify the social activities and processes that are contributory causes of these emissions, and the driving forces behind them. Who consumes which fossil fuels for what purposes? What costs and technologies are involved? What groups of actors, under which economic, political and cultural conditions, are responsible for the different patterns of worldwide soil erosion? Questions of this kind could be formulated for all core problems of global change and would constitute an impressive research agenda for the humanities and the social and behavioral sciences (the "human sciences").

Consideration must be given here to the fact that behavior relevant to global change occurs at different "aggregation levels" within society, from the individual and the family to the company and national/international organizations. Human actors occupy different roles and functions that influence their behavior in each case. People always act in spatially and temporally specific, *local* contexts, however, and are influenced by these, too – it is not the "ozone hole in itself" that moves us to buy CFC-free refrigerators, nor is it the Montreal Protocol, but reports in the daily newspaper, the example set by neighbors or fellow workers, the availability of the relevant equipment at affordable prices in the local store, the anticipated waste disposal charges for equipment containing CFCs, fear of skin cancer, etc.

Virtually all the disciplines within the human sciences study human behavior – its varieties, determinants and effects – from their specific perspectives. *Psychology*, for example, analyzes the behavior of individuals and tries to describe and explain that behavior in terms of the individual's roles and positions in society, whereas *sociology* is interested in the behavior of groups and whole societies, and therefore focuses on larger social aggregations. *Media and communications* research investigates the conveyance of information and its effects on human behavior. *Educational science* focuses on the practical options for modifying human behavior in all manner of contexts, whereas *philosophy* and *theology* reflect on normative principles and develop guiding principles for human action. Research into the political structures and institutions created by human societies is the task of *political science (see Section B 3.7)*. *Economics (see Section B 3.6)* and *law (see Section B 3.7)* study the political and economic aspects of human behavior and the conditional framework established by the legal system. *History* enables comparisons of human behavior across time, while *cultural anthropology*, *cultural geography* and *social anthropology* compare cultures.

The list above may be incomplete and contain only rough descriptions of the disciplines, but it illustrates the extraordinary breadth of the research field called the "psychosocial sphere"; under the premise that human behavior is *the* central unit for the analysis of global change, this field of research covers most of what we call the human dimensions of global change. Considering the breadth of the field, the amount of research on global change topics accomplished so far by the human sciences is astonishingly low at first glance.

One obvious reason for the dearth of global change research among the human sciences is the fact that psychosocial research, by its very nature, does not address the global dimension; the focus of research is neither global in the spatial sense, nor are time series available that correspond to the time scales of global environmental change. On the other hand, much of the environment-related research in the separate disciplines relates directly to global change phenomena, even when there is no explicit reference to these. For example, in the late 1970s and early 1980s, when there was little or no mention of the greenhouse effect, research into the psychosocial determinants of energy-saving behavior was started in response to the "oil crises" raging at the time. Such interest cannot detract, however, from the minor role traditionally played by environmental issues within wider sections of the human sciences. Very few of the theories, models and concepts commonly accepted by the various disciplines were developed in connection with environmental problems. Many of these approaches could be transferred to the global change research field without major adaptations, but the possibilities of doing so have not been subjected to broader review to this day.

Real actors – their motives and convictions, and the social barriers and conflicts between them – have played a minimal role in research explicitly addressing the human dimensions of global change. Possible reasons for this are to be found in the different theoretical foundations and premises of disciplines in the humanities, the social and behavioral sciences and natural science, secondly in the research fields on which they have traditionally focused. Research on global environmental change is dominated, due to historical factors, by a modeling paradigm based on systems analysis and natural scientific thinking. Because the human sciences have a much wider and above all more heterogeneous stock of theories and concepts compared to the natural sciences, integrating them into such models generates major problems. This heterogeneity of concepts is not in itself a drawback, but arises through the very complexity of the research object itself, which admits of several competing perspectives. But it hampers discourse within and between the disciplines, as well as the integration of relevant knowledge into the prevailing models of global change.

3.8.2
Major Contributions by German Human Science Researchers

In its report on environmental research in Germany, the German Science Council summarizes by corroborating the "... lagging development in the humanities, the behavioral and social sciences. ... The human sciences address environmental topics inadequately, and the situation is similar as regards collaboration between the human sciences and the natural sciences. The human sciences have shown little involvement in national and international research networks for the study of global aspects" (Wissenschaftsrat, 1994). This analysis is particularly pertinent to the study of global change issues, although there are definite differences between the various disciplines. Environmental research within the German human sciences and referring explicitly to global change has mainly been initiated and carried out by single disciplines, at decentralized level and on a small scale, by individual researchers and research groups at the relevant university institutes.

The DFG priority program on Human Dimensions of Global Environmental Change, which commenced in 1994 and involves research groups in psychology, sociology, geography, social anthropology, politics and economics, is a first step towards more interdisciplinarity. Supported projects cover "Perception and Assessment of Critical Global Environmental Changes and Related Behavior", "The Causes and Management of Global Environmental Problems – Analyzing Political and Economic Aspects" and "Analysis and Comparison of Resource Use Strategies in Endangered Ecosystems in the Third World".

Most of the environmental research projects within the human sciences at present are devoted to basic research and to national or local issues. The focal point of environmental sociology, for example, is the study of social structures and processes in industrialized nations. However, work of this kind can prove relevant for global change research insofar as global environmental changes are primarily induced by activities at local level.

Due to the number of disciplines and fields in the human sciences, we shall not be attempting a comprehensive review of global change research in this area. The intention instead is to identify particular topics that have a bearing on global change and are currently being studied in Germany.

3.8.2.1
Fundamental Aspects

Humanity's relationship to nature has been the subject of intense research within the field of environmental ethics in Germany since the early 1980s. Basic research of this kind centers on identifying the norms, values and motivational factors that determine or should determine people's behavior towards their natural environment. Attention focuses on two key aspects:

- *The concept of nature.* Particularly worthy of note are the investigations into what different scientific disciplines understand by "nature" and how this understanding relates to their respective methodologies. These studies provide valuable insight into the different approaches taken by scientific disciplines, and to the problems of understanding that emanate from these differences. In addition to philosophical analyses of the concept of nature, empirical approaches are applied in sociological and psychological research on values with respect to different notions of nature, the values associated with these notions and their relationship to judgments and action.
- *Legitimatory bases for environmentally sound behavior.* The philosophical debate on the legitimatory bases for environmentally sound behavior is mainly conducted between opposing *anthropocentric* and *biocentric* approaches. The central issue is whether environmentally sound behavior is substantiated on grounds of human self-interest (anthropocentric/egocentric argumentation), or by the intrinsic value of nature (patho-, bio- or physiocentric approach).

3.8.2.2
Guiding Principles of Sustainable Development

Developing *guiding principles of sustainable development* involves articulating visions for the future of society and identifying the implications these have for action in the present. Such approaches are pursued in Germany in the field of environmental ethics, for example. Sociology conducts empirical research into the conditions which favor the generation of guiding principles in society (e.g. in the study of sustainable development as a discourse phenomenon).

The Council considers the formulation of such guiding principles to be an important task, because this will trigger discourse not only amongst the academic community, but also in society at large, thus indirectly encouraging potential changes in human behavior. The underlying thesis is that technological, economic and legal frameworks are insufficient in themselves for surmounting environmental problems, but that ethical considerations and value orientations mostly give momentum to environmentally sound behavior. The study entitled "Sustainable Germany" (BUND and Misereor, 1996) includes a kind of vision of how a "high quality" of life can still be achieved while using resources in a sustainable way. The decisive aspect is not so much whether each of the recommended steps is actually achieved, but that an impulse is given for the formation of environmental awareness and appropriate value orientations.

The *implications for action* derived from environmental ethics have been formulated in Germany primarily in connection with the treatment of animals (animal ethics), with ethical issues in medicine and the natural sciences (bioethics), and with the management of technological risks (technology ethics). It is both possible and necessary to expand the range of topics to specific ethical problems of global change (e.g. protection of species).

3.8.2.3
Determinants of Human Behavior

A problem-solving approach to the modification of non-sustainable behavior can (and must) take recourse to basic research findings obtained from contexts that generally have little or nothing to do with global change. Relevant disciplines here include psychology, sociology, education, economics and law, as well as social and cultural anthropology and cultural geography.

Research on environmentally relevant behavior aims primarily at identifying and describing the determinants of those human behavioral patterns which contribute to and which help overcome existing environmental problems, as well as the interactions between these diverse factors. What, for example, leads people to travel to work by car rather than by public transport? Under what conditions is it possible to reduce the amount of drinking water consumed by households? Why does the existing level of environmental awareness have such a minimal influence on the way people act? The results of such research, which have had little explicit reference to global change phenomena until now, are often theoretical and/or empirically based multi-component models of environmental behavior (WBGU, 1996). In their heterogeneous totality, the only conclusion they permit is that a whole series of mutually interacting factors are responsible for environmentally relevant behavior, including:

- Knowledge about ecological interrelationships.
- Individual attitudes and value orientations ("environmental awareness").

- Incentives and motivations to act.
- Opportunities and offers to act.
- Factors in the cultural and social context (e.g. perceived social norms, other people acting as models to follow, existing lifestyles).
- Sociodemographic variables (age, gender, level of educational attainment, etc.).

Global change phenomena (e.g. the ozone hole or the decline of biological diversity) can rarely be experienced in a direct way by individuals, but are usually communicated by the media instead. These phenomena are complex in nature, and there is uncertainty about their real, long-term impacts. Given these specific situational aspects, research in the human sciences on the factors influencing behavior is concentrated above all on the *perception and assessment* of global change phenomena and the *cognitive processing* of relevant information (seeking, reception and processing of information).

This research field has been dominated until now by unidisciplinary approaches. One consequence is that semantic and monetary methods for evaluating environmental assets and problems are applied in isolation from each other. In addition, there has been no attempt so far to integrate the results of empirical research on attitudes and values, on the one hand, and qualitative ethical research, on the other, or research on monetary and non-monetary incentives.

SPECIFIC CONTEXTS FOR ACTION
An important research paradigm in psychology (in economics, too, but largely independent of each other) involves modeling prototypical *socio-ecological dilemmas* to analyze how people treat public goods. The structural dilemmas characteristic of global change phenomena lend a special relevance to the study of how human beings behave under corresponding conditions. Many such studies use computer simulations or experimental setups to analyze how the interdependent actions of groups depend on variations in the basic situation. One question asked is under what circumstances people's own actions are influenced by the perceived behavior of other people (e.g. long-distance air travel), its quantitative dimensions (e.g. 80% of the national population) and information about the finiteness of resources (in this example, the limits to the atmosphere's capacity to absorb pollutants). The normative dimension of the way in which the public goods problem is handled is addressed, for example, in biocentric approaches in environmental ethics which seek to grant nature its own legal status.

Specific aspects of the global change problem are also touched on in research looking at how people deal with *complex situations* or problems. Such studies have shown that people's capacity for "systemic thinking" is limited. When faced by complex problems, people will tend to fall back on established, simplifying and often inappropriate patterns of thinking and problem-solving.

A whole series of mostly empirical studies has been carried out into how people act under conditions of *uncertainty*, how *risks* are handled, and the respective factors which operate. Scientists in the fields of psychology, sociology and communications research, in particular, are concerned with perception, acceptance and communication of risks. However, the relevant research is mostly concentrated on the risks associated with large-scale technologies (e.g. nuclear energy, biotechnology and genetic engineering). Global change phenomena, in contrast, have been clearly neglected. The same shortcomings are apparent in ethics research on these topics, which focuses on the risks that society can reasonably be expected to tolerate, improving risk analysis methods, ethical and legal implications of risk assessments, and the conditions for a collectively binding, rational comparison of risks.

SOCIETAL ACTORS

Human science research aimed at solving global change problems has no alternative but to focus on individual actors and groups of actors and to analyze their behavior at different levels of society. This includes investigating environmentally relevant behavior and its determinants in specific societal *subgroups* (e.g. women, children, the elderly), as well as the implications of the various *roles* that individuals assume in life (e.g. blue-collar workers, industrialists, politicians, multipliers). German research has addressed these questions to only a marginal extent so far, e.g. in studies on environmental awareness among industrial workers, or gender-related differences in environmental behavior.

SPECIFIC PATTERNS OF BEHAVIOR

Research in Germany on environmentally relevant human behavior and its determinants is concentrated primarily on certain areas of consumer behavior. For example, there are numerous studies on how households handle energy (energy-saving behavior) and waste (separation and avoidance of domestic waste); more recently, there has been growing interest in the ways that people use and manage water. Another, relatively common topic of research is personal mobility, and choice of transport especially. Tourism and leisure research is another related area. Relatively little effort has been dedicated to the integrative study of different behavioral patterns using more comprehensive concepts (e.g. lifestyles).

STRATEGIES FOR BEHAVIOR MODIFICATION

The debate on adequate instruments of environmental policy is dominated by the two poles of "command and control" and "economic instruments". Psychological and educational strategies for influencing human behavior are often subsumed as a somewhat exotic residual category under the short but misleading term "moral suasion", and are often reduced in the process to the mere purveying of information. There may be historical reasons for this somewhat straitjacketed conduct of the environmental policy debate, but it shows up deficits in the way that the human sciences study behavior modification strategies. The latter has been characterized so far by a more experimental approach (small samples, often at the level of private households) and by the consideration given in each case to only a few factors potentially influencing behavior. Although generalizable research results are now available for this level, the interaction between the various determinants of behavior, for example in *environmental education* programs, is largely unresearched, as the Council has already pointed out in a previous Report (WBGU, 1996). Moreover, there are no studies that permit an assessment of the interactions between individual strategies in environmental policy, for example by integrating these strategies within a common conceptual framework. Regular interdisciplinary discourse is obviously lacking among all those disciplines which share the common aim of modifying environmentally harmful behavior – psychology, sociology and economics in particular.

ENVIRONMENTAL DISCOURSE

A more recent field that communications research shares with sociology and political science is the analysis of social discourse on environmental issues. One important theme here is the *analysis of the formation and development* of such discourse. The climate of public opinion has direct relevance for decision-making processes in the political, business and private spheres. The public "communications arena", the main component of which is media reporting, is where issues are defined and structured, and global change is no exception. Knowledge about the conditions which shape the formation and development of public opinion on environmental issues can thus be used to derive policy recommendations. Environmental discourse is also the overarching topic of sociological research on the origins of the environmentalist movement, and of psychological and sociological research on mediated negotiations (e.g. mediation processes, *see Section B 3.7.4.4*).

3.8.3
Integration of German Human Science Research in International Programs

In view of the deficits apparent in German research on global change topics, it comes as little surprise that current involvement in international programs such as the International Human Dimensions of Global Environmental Change Programme (IHDP, *see Section B 1.3*) is low. Having said that, the current Chairperson of the IHDP Steering Committee is a German, and a national HDP Committee is currently being set up. Explicit reference to IHDP is also made in the application to establish the DFG priority program on Human Dimensions of Global Environmental Change. Explicit "identification" with the IHDP program cannot be made out as yet, however, neither at project nor at program level, so an assessment of the specific contributions of the priority program to IHDP is not possible at this point in time. Representatives of the priority program are active on the Steering Committee of the new ESF program entitled Tackling Environmental Resource Management (TERM, *see Box 5*).

The EU's Fourth Framework Programme for Research and Technological Development, which features a predominantly natural scientific and technological orientation, includes support programs on sociocultural and socioeconomic aspects of global change (*see Box 5*). In the "Environment and Climate" program, for example, funding is provided for research projects on "The Human Dimension of Environmental Change". In the Targeted Socioeconomic Research activities (TSER), support is given to global change research under the activities "Evaluation of science and technology policy options for Europe" and "Research into social integration and social exclusion in Europe". The amount of research funding disbursed to human science researchers from Germany under this program is not known.

3.8.4
German Research in the Human Sciences – Research Needs Concerning Global Change

In general, global change research in the human sciences must become *more policy-oriented*. For example, the various disciplines could examine the issues negotiated in connection with global change conventions with respect to their social and political feasibility, their culturally specific acceptance and social compatibility. Thus, potential communications problems and barriers to action due to different value systems, attitudes and behavioral patterns and so-

cioeconomic conditions, could be rendered transparent in the run-up to negotiations. Furthermore, a more policy-oriented approach on the part of research must also involve the systematic analysis and evaluation of all measures aimed at sustainable development. A greater *orientation towards research application and problem-solving* can be achieved by gearing psychosocial research more towards identifying behavioral determinants and strategies for behavior modification at all levels of individual, social and institutional action. An interesting field is opening up here for basic and applied research.

Since global environmental problems are primarily the consequence of local behaviors, it is necessary in terms of *methodology* to study actors and groups of actors in the respective contexts within which they act. Particularly important in this connection is culturally specific and cross-cultural research on social groups, in the form of comprehensive and transdisciplinary case studies. Proceeding on this basis, what is then recommended is a widening of the temporal and spatial horizons of research, i.e. greater integration of temporal dynamics (e.g. through longitudinal studies), as well as the creation of larger spatial reference frames in the form of large-scale study areas, within the framework of comparative transcultural studies and through an explicit focus on global problems.

The systemic nature of global change virtually compels the research community to engage in dialog and *interdisciplinary or transdisciplinary collaboration*, both within the human sciences and between the natural and human sciences. The rigid structures of university funding, the appointments system and the associated lack of career opportunities for scientists with an interdisciplinary orientation have been detrimental for such dialog. These factors are compounded by the review procedures of the principal research support bodies (*see Section C 8*), which are mainly organized according to discrete disciplines. The question as to how an adequate form of interdisciplinary cooperation can be achieved is raised – not least on account of methodological differences – both within and between the human and natural sciences. The evaluation of existing and the development of new, integrative research methods are key challenges for global change research:

- There is already a considerable body of research results from individual disciplines which explicitly address or are related in some way to global change issues (*see Section B 3.8.2*). To utilize this knowledge more effectively, it is necessary to develop instruments for combining, integrating and evaluating *existing* research results in order to solve specific problems (*ex post integration*).
- At the same time, integrative approaches must be developed that enable interdisciplinary research

on global change *from the very outset* and which are geared to the study of ecosphere-anthroposphere interactions (*ex ante integration*). In addition to the search for possible alternatives to the leading paradigm of environmental research, systems analysis, this should also include the study of existing integrative research approaches and models with respect to their premises, implicit value judgments, and the origins of theories, data and predictions, with consideration also given to aspects of the theory of knowledge and science. It is particularly important to unveil the "human dimension" in the prevailing approaches for modeling economic and social systems, and to develop research approaches that permit reference to real actors and groups in their specific spatial and sociocultural contexts. Only in this way is it possible to achieve greater integration of the human sciences into the wider field of global change research, which has been dominated so far by the natural sciences.

3.8.4.1
Scientific Issues

The precondition for a more intensive involvement of the human sciences in global change research, besides the transfer of existing theories to global change issues, is the development of new social scientific concepts of globality. Possible starting points include the advancing globalization of economic activities, increasing individualization accompanied by cultural homogenization of global society ("McDonaldization"), global information networks, and the population growth, urbanization and migration complex.

Research has focused on national and local issues until now, and now needs to widen its approach with a global perspective (e.g. comparative cross-cultural studies, analyses of regional and global contexts) and a greater orientation to the policymaking process and the ensuing research needs.

GUIDING PRINCIPLES OF SUSTAINABLE DEVELOPMENT

The Council considers it essential, in the context of efforts to tackle environmental problems by means of international conventions, to explore the ethical implications of these political processes. This applies in particular to AGENDA 21 and the conventions agreed in the aftermath of Rio, which have mostly been analyzed in terms of natural scientific, political and economic aspects, but little from the humanities and social science perspective. The Council

therefore recommends that support be given to research projects on the following topics:

- Implications of AGENDA 21, the global conventions (Climate, Biodiversity and Desertification Conventions) and the concept of sustainable development for social value systems and the legal, economic and educational strategies which ensue (culturally specific and cross-cultural research).
- Social acceptance and social compatibility of the terrain covered by the conventions (convention-related research in the fields of social and behavioral science and ethics).

Besides these activities, it is also important to promote the interdisciplinary development of culturally specific, qualitative visions (guiding principles) for an environmentally sound way of life.

DETERMINANTS OF HUMAN BEHAVIOR

Suitable programs aimed at modifying behavioral patterns of relevance to global change cannot be designed without an improved understanding of the factors that influence such behavior and unless the relative significance of these factors have been investigated. Global change, in all its facets, is characterized above all by the lack of "tangibility" and by extensive uncertainty regarding its precise effects. For this reason, the study of the *perception and evaluation* of global change phenomena and their relevance for action represent a major challenge for the human sciences. Descriptive and diagnostic approaches must be supplemented here by explanatory studies in order to generate an improved understanding of how corresponding information processing systems operate. The Council's recommendations in this respect are:

- Development and establishment of a worldwide, comprehensive social monitoring system for continuous, comparative and descriptive analysis of perceptions, attitudes, motivations and behaviors of relevance to global change at different levels of social aggregation (potential starting point: IHDP-GOES, *see Section B 1.3*).
- Study of the cognitive, emotional and motivational processes that operate when information on global change is processed by individuals.
- Combination of existing and development of new interdisciplinary approaches for the evaluation of environmental assets and global change phenomena and for setting policy priorities, with specific reference to social, behavioral and ethical aspects.
- Combination of results from empirical values research (especially in psychology and sociology) and from normative-ethical approaches.

Specific Contexts for Action

Human science research into the determinants of human behavior in *socio-ecological dilemmas* should be carried out more intensively from the global change perspective in future. There is considerable potential for interdisciplinary cooperation here. The problem of public goods is another area where the Council recommends increased collaboration between ethical philosophers, political scientists, legal experts, economists and psychologists.

Given the globality of the environmental changes now being observed, it is also essential to stimulate the integration of sociological, political and ethical issues into the debate on *equity aspects* of the worldwide distribution of environmental stress, on the one hand, and environmental protection measures, on the other (*see Section B 3.7.5.2*). This debate has been conducted in predominantly economic terms until now.

Existing research on the *perception and acceptance of risks* should be extended to cover global change topics and to acquire a greater interdisciplinary focus. A possible focal point could be the influence exerted by the perception of risks emanating from global change phenomena on the acceptance of such risks, on decision making processes and on environmentally relevant behavior in practice.

Societal Actors

More than hitherto, human science research should identify *real actors* and groups of actors within global change and investigate them in their respective behavioral contexts. Relevant groups that have not been sufficiently studied until now include, for example, decision makers in the political and business domains, and multipliers (journalists, for example). This would be beneficial in designing target-group strategies for modifying behavior, which as a rule are highly effective and cost-efficient.

Specific Patterns of Behavior

In the last analysis, it is not so much individual and distinct elements of behavior (energy and water saving, mobility, etc.) that are non-sustainable and which contribute significantly to global environment problems, but the production methods and consumption patterns of the "developed" countries as a whole, and the *lifestyles* they entail. To develop alternatives for these, it is necessary to identify and investigate complex behavioral patterns in the relevant cultures, their values and norms and their economic systems, and the scope for action arising from these factors. Integrative research approaches provide an appropriate vehicle for such analysis, beyond the confines of the individual disciplines.

Furthermore, the different stages and paths of development in global society should be described with respect to the attitudes towards and the use of natural resources, and subjected to comparative analysis. In the course of time, it should be possible to pinpoint possible determinants of sustainable development or environmentally more compatible ways of using natural resources.

Nevertheless, research on the causes and effects of *individual* modes of behavior and their interlinkages continues to be necessary, especially against the cultural background of the developing countries. Possible topics include:

- migration from threatened areas.
- urbanization and waste management.
- tourism and global change.
- local self-help in the informal sector.
- sociocultural factors influencing reproductive behavior.

Strategies for Behavior Modification

The environmental policy instruments developed above all by political science, law, economics and psychology and aimed at *changing behavior* at different levels of society still tend to be analyzed in isolation from each other. Research into ways of integrating these various instruments should be intensified, since this could lead to an improved understanding of the driving forces of human behavior, as well as synergies in the management of global environmental changes. To determine whether existing behavioral patterns are coming into line with new guiding principles and goals, all environmental policy instruments must be subjected to regular evaluation of their effectiveness.

It is now apparent that, if environmentally benign behavior is to be disseminated among the population, the entire spectrum of factors which may influence behavior has to be taken into consideration and adapted if necessary to the respective target groups and contexts. Therefore, more case studies need to be conducted in order to ascertain the specific conditional frameworks under which particular combinations of interventional methods will be effective.

Environmental Discourse

The *media* play a special role in the formation of public opinion on global environmental problems. Specific aspects of communication on global change are therefore in need of analysis within an interdisciplinary research framework oriented to practical applications and solutions. Such research must investigate the diverse factors on which the perception and description of problems depend; more efficient *communications strategies* must then be elaborated on this basis, in the following areas especially:

- Factors involved in focusing and binding public attention (agenda setting).
- Role of public communications system in the incidence, conduct and settlement of environmental controversies.
- Determinants of decision-making and behaviorally relevant processes of information dissemination and communication in the public, political and business spheres.
- Supply and infrastructure of information on global change, and factors preventing their being used.
- Factors affecting the development of supranational "publics" and collective representation.

At a more general level, research results on environmental discourse obtained primarily within the individual disciplines should be integrated through interdisciplinary collaboration wherever overlaps occur (*see Section 3.7.4.4*).

3.8.4.2
Structural Improvements Needed

The global environmental crisis is caused by human activities, and is therefore rooted in a crisis of society. This fact, although commonly accepted in society itself in the meantime, is poorly reflected in the German research landscape. The first requirement is therefore a significant increase in the support provided for human science research on global change, whereby the goal must be greater institutionalization of this research field and the evolution of corresponding perspectives on global change phenomena. Global change research within the human sciences in Germany has been weakly organized until now, due primarily to the dominance of single-discipline approaches. The universities and research support bodies must respond accordingly in order to create the conditions for networking research within and between the disciplines. One starting point for the promotion of interdisciplinary research activities would be the formation of temporary research groups, as has already been recommended by the German Science Council (Wissenschaftsrat, 1994).

Global environmental changes are caused and influenced by local activities to a major extent. The specific context within which people act is therefore of major significance, so it is absolutely essential that greater efforts be invested in the development of national human dimensions programs. At the same time, certain issues can only be tackled in a meaningful way through international cooperation. Examples include cross-cultural research and the establishment of a comprehensive worldwide network for social monitoring, analogous to the environmental monitoring networks already in place. Further expansion of international human dimensions programs on global change, especially the IHDP, is therefore necessary. In view of the overriding importance of the human sciences for the *behavioral* problem of global change, international programs with real "clout" would broadcast important signals to those involved in national and international politics to give greater consideration to psychosocial aspects within particular problem-solving approaches. The research agenda of IHDP and the specific projects envisaged within the program (LUCC, GOES, IHDP-DIS, START) provide a number of concrete areas where German involvement could be stepped up (*see Section B 1.3*).

3.9
Technological Research

3.9.1
Relevance of Technology for Global Change

Technological research centering on the problems of global change is primarily aimed at finding improved or new environmentally sound technologies for achieving sustainable development, especially for sustainable preservation of life-support systems. Creating a harmonious link between environmental protection and economic development (AGENDA 21, Chap. 31, Section 8) requires enhanced research and development efforts relating to environmentally sound technologies and their socioeconomic interactions. Technological research and development should also provide the vital support needed in order to manage and avoid the harmful impacts of global change, and must be integrated in the search for holistic solutions (WBGU, 1993). This includes the compilation of criteria for the assessment of new technologies with respect to their impacts on all areas of the environment and on human health.

Energy research focusing on climate protection is of special importance in this context. Such research must also integrate work on air pollution prevention, in that the latter generates emission reduction technologies for substances that contribute to the formation of tropospheric ozone. "Substance-based life cycle management" is a relatively new research field that could lead worldwide to a departure from open materials management systems.

3.9.2
Major Contributions by German Technological Research

Environmental engineering research in Germany is conducted at a high standard, covering all the key environmental media and fields: waste management, remediation of contaminated sites, air pollution prevention, low-emission technologies, water protection, water supply and sewage disposal, noise abatement, as well as safety technologies for high-risk installations, systems and services. Details are contained in the various support programs of the German Federal Government (Panzer, 1995).

In terms of relevance to global change, special attention must be directed to the renewable energy sources examined by priority research in the energy field, i.e. hydroelectric, wind, solar and biomass energy. In this area, too, new technological solutions have been found or are being worked on. Special mention must also be made of research on Rational Use of Energy (RUA), which is leading to the implementation of many technological innovations. Details are provided in the Global Environmental Change 1992-1995 research framework established by the BMFT (April 1992) and the final report of the Enquete Commission Protecting the Earth's Atmosphere set up by the 12th German Bundestag (1995a).

search and development work carried out within the framework of the energy research programs of Germany and the EU is devoted to the reduction of energy consumption. Research commissioned and supported by the EU explicitly requires international cooperation on the part of at least two European institutions. Environment and Climate 1995-1998, an EU research and development program, provides support to RTD activities in the field of environmental technologies; *Box 11* provides an overview.

The NATO Committee on the Challenges of Modern Society is funding a special program for the remediation of contaminated sites, with participant researchers from Germany and nine other nations. Bilateral cooperation with developing and newly industrializing countries is also important, and has been greatly strengthened in recent years. An even greater fusion of capacities should be engineered in future at both national and international level to ensure that synergy effects are achieved despite cut-backs in funding. Due to the intense international competition in the environmental technology sector, endeavors must be made to bolster international collaboration in the field of pre-competitive research (Enquete Commission, 1995a). If synergies between different approaches are to be exploited, particularly in energy research, then cooperation with international organizations has to be intensified (e.g. International Energy Agency, IEA).

3.9.3
Integration of German Technological Research in International Programs

Another key requirement for solving global environmental problems is international collaboration in the field of technology research. Much of the re-

3.9.4
German Technological Research – Research Needs Concerning Global Change

Further development of currently available technologies and the development of new environmentally sound technologies for the prevention and miti-

---

BOX 11

Main Activities of the EU Research and Development Program Environment and Climate 1995-1998 (excerpt)

1 Instruments, technology and methods for monitoring the environment
2 Technologies and methods for evaluating environmental risks and for protecting and restoring the environment
  2.1 Methods for evaluating and managing risks for humankind and the environment
  2.2 Analysis of the life cycles of industrial and synthetic products
  2.3 Technologies for protecting and restoring the environment
    2.3.1 Clean technologies and clean products
    2.3.2 Emission-reducing technologies
    2.3.3 Recycling technologies
    2.3.4 Organic wastes
    2.3.5 Hazardous wastes
    2.3.6 Remediation of contaminated sites

Source: European Commission, 1994

gation of global environmental changes should continue to play an important role in the Federal Government's research portfolio. This calls for promotion of technological development in virtually every field, with priority attached to significant reductions in the consumption of energy and materials in the extraction of natural resources, in the production, distribution, consumption and disposal of goods, and in the service industries.

3.9.4.1
Climate Protection Technologies

In the view of the Council, special emphasis must be placed on researching and developing improved technologies for climate protection, particularly in the energy sector. Germany's share of the rapidly growing market for energy technologies is currently about 20%. This means that German technology has a significant influence on global developments in which energy factors play a role. The Council recommends that, from the wide range of technological research fields, the following topics be selected as priority areas.

RATIONAL CONVERSION AND USE OF ENERGY RESOURCES
New and further development of conversion, end-use and useful energies for *rational conversion of energy resources*, including basic research on high-temperature material physics and exergy thermodynamics, should be given top priority. The emissions generated by present-day systems for converting and using energy are by far the greatest source of climate-forcing trace gases in the industrialized nations (Enquete Commission, 1995a). However, quantitative predictions of technology-based reduction potentials involve considerable uncertainties. The Council is convinced of the enormous potential which could be exploited here, not only for industrialized countries, but also in the context of industrial development in newly industrializing and developing countries.

The Council sees great promise in current efforts to refine the *process integration method* for higher energy efficiency in industry (*see also* UBA, 1994). Implementation and broad application of this method in German industry requires a larger number of case studies covering as many sectors as possible. This should be followed up by international programs for the exchange and transfer of know-how and technology aimed at enabling industrial enterprises in the newly industrializing and developing nations to adapt accordingly.

In many cases, the energy-saving components and systems spawned by research and development activities have to be integrated into existing facilities and equipment. Such adaptation calls for engineering expertise that is able to find ways and means of systems optimization. The Council recommends that projects of this kind be included in the government's research program to a greater extent than hitherto. Furthermore, it is essential to improve conditions for medium-sized German companies so that they can actively take part in the international energy-saving project organized by the IEA.

Research and development activities in Germany should maintain the current focus on small-scale consumption and households (e.g. saving energy). However, these projects should be supported by ongoing projects in the field of information and training, in order to dismantle social barriers and tap the energy-saving potential in the private consumption sector.

RENEWABLE ENERGIES
The overall aim in this field should be to provide continuous R&D support and to test interesting new approaches in connection with heating, electricity and renewable fuel consumption, in order to effect the necessary changes in the energy mix, particularly with respect to global environmental change.

Photovoltaic systems have by far the greatest potential worldwide for power generation using renewable energies. A long-term approach within the framework of a research program entitled "Photovoltaics by the Year 2020" will be necessary to achieve the required cost reductions and desired market success for this technology. The top priorities of this program should be to advance photovoltaics technology based on crystalline silicon, and to cover the following areas (for further details *see Box 12*):
- semiconductor technology, including thin-film solar arrays;
- systems structure technology, including modular links, power electronics, switchgear and safety systems;
- production systems for more efficient series manufacture of fully developed products.

It would be advantageous here to draw on the experience gained during the support program on "Measurement and Documentation at 49 PV Installations", which has been running since 1988 (MuD et al., 1996).

Technology-based measures should be accompanied by a broadly based program for market development. A package of activities aimed at markets inside and outside Europe is essential if, through growth in market volume, the necessary and possible cost-degression mechanisms are to operate. Such a program requires a socioeconomic approach. A careful analysis must be made of the factors that have prevented widespread application in developing countries (such

---

BOX 12

Research Topics in the Field of Solar Energy Systems

- Further development of thin-film technology: demonstration of energy production technologies in the 100 kW/a to 1 MW range in order to generate reliable projections of costs, including optimization of environmental soundness of production processes and recycling techniques. Development of thin-film modules for mass production, especially modules for buildings.
- Development and demonstration of production technologies for large-scale modules using amorphous silicon and featuring high and stable efficiencies.
- Applications-oriented fundamental research to develop solar cells based on new materials.

- Further development in the field of electro-technical systems engineering adapted to photovoltaics: power electronics, load management systems, monitoring systems, electromagnetic compatibility.
- Development and demonstration of optimized techniques for integrating photovoltaic systems into buildings, settlements and large building complexes: standardization, higher levels of system efficiency, elaboration of service concepts.
- Development and demonstration of photovoltaic energy supply concepts in developing countries, including quality assurance, training, financing and local production of system components.

Sources: Kleinkauf et al., 1993, DECHEMA, 1994; Enquete Commission, 1995a; Luther, personal communication, 1996;

---

as cultural influences), and new solutions found for improved adaptation.

In spite of the increasing use of *wind energy* in Germany, it is still necessary to ensure continuity in the research and development field. This is particularly the case in respect of land-use reductions, dismantling barriers to acceptance by appropriate siting of turbines, and noise abatement methods. To boost export-oriented development and market development, a research program should be set up to solve the specific problems associated with installation, integration into existing supply structures, and maintenance (remote diagnosis) of systems abroad (particularly in developing countries).

Marketing renewable energies often requires research support. Funding to this end should be provided for developing countries not only by the Federal Government, but also by international organizations, especially the World Bank and regional development banks (e.g. Asian Development Bank). The Council welcomes the incentive program for renewable energies initiated by the Federal Government (1995 to 1998), which also supports biomass technologies.

STORAGE AND TRANSMISSION TECHNOLOGIES

Storage systems of all kinds, especially for heat and electricity storage, as well as chemical and electrochemical systems, should be developed intensively, with consideration given not only to technical, but also to ecological aspects. Heat and power generation using renewable energies is subject to temporal fluc-

tuations, which means that more economically efficient storage systems are urgently needed. In the field of power transmission, further research into the possibilities offered by superconductivity should be carried out within a worldwide network.

POWER GENERATION TECHNOLOGIES

Improvements in power generation technology necessitate continuous research on combustion processes and the transformations of materials involved in such processes. This is research of a fundamental nature and should not be neglected. To integrate high-temperature fuel cells into the gas and steam turbine process, more research is necessary on materials and fuel cell production (Enquete Commission, 1995a). Research aimed at further development of fuel cells must also be seen in the context of developing Carnot-independent electrochemical energy conversion processes. Further applications-oriented work would also be useful for unit-type district heating power stations based on fuel cells, given the fact that these exhibit a significantly better power/heat ratio than combustion engines (Luther, 1996). Another important field is the refinement of fuel cell applications for motor vehicles (zero emission car).

There have been definite advances towards higher efficiency levels in power generation technology and in the development of new power station concepts based on fossil fuels. Some of these advances have now been put into operational practice. Demonstration projects, involving transnational cooperation for example, should be promoted in all areas where a major contribution towards reduction of carbon dioxide emissions can be expected. The Council regards the "Investments for Reducing Environmental Stress" and "Investments for Reducing Transboundary Environmental Stress" programs set up by the BMU as a useful means for promoting pilot projects aimed at disseminating advanced technologies of relevance to global change to other countries.

AIR TRAFFIC TECHNOLOGIES

Increasing air traffic (WBGU, 1993) is becoming an increasingly important source of greenhouse gas emissions (*see Mass Tourism Syndrome, Section C 2.2.1*). It is therefore essential to continue research into the global impacts of these emissions. The development of low-emission aircraft engines has now been included in the joint "Pollutants in Aviation" program that has been running since 1992. Another issue which should be examined is whether it will be necessary in the long run to develop new types of climate-compatible aircraft. The specific requirements would relate in particular to fuel consumption, speed limitations and flying height (Enquete Commission, 1995b). The technical innovations (e.g. use of hydrogen) should be accomplished through worldwide cooperation, integrating the technological advances achieved in Russia.

3.9.4.2
Technologies for Protecting the Ozone Layer

Further research is needed to determine the technological potential for reducing anthropogenic changes in atmospheric ozone concentrations. The changeover to substitutes for CFCs and halons must be completed more rapidly by promoting research and development work, and coordinated internationally (especially for developing nations). In the long term, anthropogenic N_2O emissions account for an increasing proportion of all emissions, so all sources of N_2O emissions must be investigated thoroughly. The development of N_2O emission reduction technologies is also essential.

3.9.4.3
Technologies Relating to Material Flows

The reports of the Enquete Commissions of the German Bundestag (Enquete Commission, 1993 and 1995 a and b) already contain numerous proposals for research and further development concerning the sustainable management of material flows. Building on the idea of ecological product line controlling

(UBA, 1994) as a basic tool for managing material flows, any analysis of the need for low-consumption technologies must cover manufacturing, energy supply and services, and take economic and social acceptability into account. In this context, the Council welcomes the Production 2000 framework concept of the German Ministry of Education and Research. Consideration should be given not only to projects in the production field, but also to those focusing on resource extraction and the service industries, whereby a global perspective should be taken as well. Management of material flows also applies to waste disposal, so additional research topics geared to international waste management are recommended (*see Box 13*).

3.9.4.4
The Technology-Economy Interface

Technological research aimed at solving global environmental problems is always bound up with economic issues, and to that extent has cross-cutting dimensions (Rentz, 1995). For this reason, the Council recommends a number of research activities at the technology-economy interface:

- Examination of the appropriateness and effectiveness of jointly implemented activities to reduce greenhouse gases.
- Development and assessment of cost-efficient reduction strategies for greenhouse gas emissions, taking the radiatively active trace gases (CO_2, CH_4, N_2O, O_3) into account.
- Exploration of techniques for removing and storing CO_2, with special reference to ecological and economic aspects.
- Quantification of the impacts of greenhouse gas reduction strategies on the emissions of other mass contaminants of the atmosphere.
- Development of cost-efficient strategies for reducing tropospheric ozone.
- Identification of environmentally sound industrialization pathways in developing and newly-industrializing countries.
- Further development of locally adapted technologies in developing countries (e.g. local transport systems and production methods, traditional building methods).
- Analysis of the influence of government measures on the development of low-emission and low-residue technologies.
- Development of combined plants at the company and industry-wide level in order to optimize the closed substance cycle economy (e.g. combining production plants to save energy and raw materials, and to avoid emissions and residuals).

BOX 13

Research on Global Waste Problems

- Elaboration of strategies to ensure that the waste produced by the growing world population is disposed of in environmentally sound ways, in accordance with a globally coordinated concept.
- Development of strategies and mechanisms enabling waste avoidance, recycling and disposal measures that apply to industrialized as well as developing countries.
- These strategies and mechanisms must be designed in such a way that participation is an attractive option for as many countries as possible. Adaptation processes must be structured in such a way that countries consider implementation of these measures to be in their own interest.
- Research priorities:
  - Analyzing the interests of individual countries or country groups.

- Identification of the avoidance, recycling and disposal phases, and the harmonies or conflicts of interest specific to each phase.

Detailed concepts for the individual country groups could then be developed on this basis. Where intervening action is planned, the concepts in question must also relate to the regional and local (municipal) level, taking the different social and settlement structures into account.

The project should be carried out in two stages:

Stage 1:

Determinants of waste avoidance, recycling and disposal; relevant modes of behavior and their modification.

Stage 2:

Planning of intervention measures, with consideration given to nationally, culturally and locally specific aspects, as well as economic, technical, legal, social and psychological factors.

The project should be carried out by an interdisciplinary working group comprising environmental psychologists, economists, engineers, and foreign experts from the various country groups.

- Development of logistics-oriented production processes (e.g. reduction of transport pathways within the production process).
- Development of production processes and technologies for efficient utilization of materials and energy.

3.9.4.5
Structural Improvements Needed

Experience with practical, technology-based environmental policy has shown that complex environmental engineering problems require cooperation between the following disciplines:

- Technologies: engineering sciences.
- Materials: chemistry, biology and geology.
- Planning and design: economics, social sciences.
- Applications and effects: social and behavioral sciences, environmental medicine.

Finding practicable solutions to complex environmental engineering problems, i.e. drawing up specific recommendations for the design of systems, requires that all of these areas be covered within each individual project or for the specific problem at hand. Mere linkages between different university institutes are not always an adequate approach. Greater benefits are achieved if researchers from a variety of fields

are brought together in a single working group, i.e. under the same roof. An assessment of the status quo reveals that multidisciplinary institutes and organizational structures exist in only a few faculties or departments in Germany. There is obvious room for improvement here, whereby lessons should be drawn from the environmental research centers and their efficiency (Rentz, 1995).

3.10
Summary: Current Status of Global Change Research in Germany

The German research community has made some key contributions to our understanding of global environmental changes. However, there are major disparities between the various scientific disciplines with respect to the organization and breadth of research activities, the extent to which they are involved in international programs, and their effective output.

Within the natural sciences, special priority is attached to the fields of climate and atmosphere research, as well as marine and polar research. Several collaborative research centers and priority programs of the DFG exist in these fields, and there is intensive involvement in international programs. In contrast, the global dimension in the fields of lithosphere, pedosphere and especially biosphere research is still relatively underdeveloped in Germany.

The humanities and the social sciences are only beginning to show an explicit interest in global change research. They tend to operate within the confines of single disciplines and exhibit a strong bias towards nationally oriented research. While this may accord with the specific objects of analysis (cultures, societies, individuals), it is absolutely imperative that they give greater consideration to the global dimension in future. The prerequisite for this is the formulation of basic concepts for studying and analyzing global change phenomena (e.g. concepts of globality), and support for additional research projects and programs alongside the only priority program of the DFG in this area.

The study of global change as a multi-facetted layered, interdependent phenomenon requires cooperation between scientists from different disciplines. Predictive models of anthropogenic environmental changes developed by the natural sciences rely on assumptions about the future behavior of people, and therefore require close cooperation between natural scientists and social scientists. Conversely, if the latter aim to formulate concepts for coping with environmental problems, they need valid data from the natural sciences. The German research community has yet to achieve this all-important networking of the social and natural sciences.

The same can generally be said of interdisciplinary research within the social sciences, and to a certain extent within the natural sciences. Interdisciplinarity of a kind has been introduced in some areas (research into forest decline being one example), but such efforts must be strengthened and multiplied in future.

A general problem affecting global change research in Germany is its lack of problem-solving competence and hence political relevance. Outstanding research by itself will not solve environmental problems. Instead, research must demonstrate an applications-oriented approach, and it is necessary to translate output into concrete political action and to formulate environmental goals in a way that facilitates practical implementation of specific objectives. This is all the more necessary if policymakers are to act according to the precautionary principle: politicians can only take active steps if research data is translated into political demands. The scientific community must support the policymaking process by orienting its research activities to real needs, in other words to the demands that ensue from ongoing political processes (e.g. negotiations on the various environmental conventions).

Global change research in Germany has been largely a preserve of the natural sciences. It lacks interdisciplinarity, international collaboration and problem-solving competence for combating acute and potential threats to the global environment. In the next main Section of this Report, the Council responds to this situation by suggesting ways in which these shortcomings can be eliminated.

New Guidelines for Environmental Research

Section B 3 of the Report focuses on the various sectors and disciplines of German global change research. As explained in the Introduction, the Council considers this sectoral approach to be inappropriate and in need of revision. The complex phenomena of global change cannot be analyzed in a purely sectoral manner or from the perspective of a single discipline, because they are the result of multilayered interactions between the ecosphere and the anthroposphere. It is virtually unthinkable that global change research could provide the basis for new response strategies without an analysis of the complex interactions between processes in the ecosphere and anthroposphere (population growth, economic trends, as well as technological and psychosocial processes).

The specific content and organization of German global change research must demonstrate such a transdisciplinary view to a greater extent than hitherto. In its various Reports, the Council has elaborated an approach that meets these criteria. This systems approach, as expressed in the *syndrome concept*, is applied step-by-step in the following chapter to derive new guidelines for global change research.

The interdependency of human and natural systems demands an approach which ensures that the complex problems of global change are analyzed in an integrated way from different perspectives and at a variety of levels. The Council believes that this integration must occur in both the horizontal and vertical dimensions. *Horizontal integration* refers to the problems themselves, their basic structures and their interconnections. *Sections C 2 to C 6* are devoted to this aspect. To generate response strategies for each problem area, it is necessary to approach them from another angle entirely. We refer here to *vertical integration*, which proceeds in various stages from policy-based analysis of a problem, via the implementation of appropriate instruments to monitoring of the latter's effectiveness (*Section C 7*).

The central focus in this part of the Report is on *horizontal integration*. The key instrument used is the *syndrome concept*, which the Council applies in this context to research activities. Syndromes are derived from the *global network of interrelations* and repre-

sent complex "clinical profiles" (*Section C 2.1*). These are the product of characteristic constellations of socioeconomic, geographical and political trends within the network of interrelations, and can be identified in many regions of the world. The Council has now made its first attempt to compile a complete list of syndromes (*Section C 2.2*), assigning the syndromes to the *core problems of global changes* (*Section C 2.3*).

It would be unreasonable to expect the research community to study all the syndromes simultaneously and comprehensively. Criteria are therefore needed for weighting syndromes and investigating them in detail with an appropriate form of research organization. The Council has developed two types of criteria to this end:

1. *Relevance criteria* serve to rank the syndromes according to their "importance" from the perspective of German research (*Section C 3*). One such criterion is existing competence within the German research community, which can enable solutions to be found relatively quickly.

2. *Integration principles* are necessary in order to translate the demand for networking or interdisciplinarity into specific requirements to be fulfilled by research programs and projects (*Section C 4*). Such principles relate to features of the research object and the relevant methodologies, as well as aspects of research organization and subsequent action in response to the findings.

Syndrome ranking (*Section C 5*) is then carried out on the basis of these principles. The Council has conducted a survey of its members in order to test the applicability of the principles. Final, improved ranking should be done by a larger group of experts, in a Delphi study, for example.

Once the most important syndromes of global change have been identified and ranked, it is then possible to design the research needed for selected syndromes. Taking the *Sahel Syndrome* as an example, we outline how the relevant research should be structured (*Section C 6*).

Researching such complex problems in order to develop and implement response strategies requires a range of research structures and funding instruments that can expand on existing setups, on the one hand, but which have yet to be developed and tested anew, on the other *(Sections C 7 and C 8)*.

2.1
The Systems Approach

A cardinal feature of global change is that humankind itself is now an active factor within the Earth System, playing a significant role at the planetary scale. Human interventions, as manifested in the depletion of raw materials, shifts in material and energy fluxes, changes to large-scale natural structures and critical stresses on environmental assets, are altering the very character of the Earth System to an increasing degree. The complexity of the processes involved or driven by these changes poses a major challenge for the scientific community and generates a number of new research issues. Finding answers to these questions will be of increasing importance in the years to come:

• What are the causes of these changes in the ecosphere and how are these linked to global development problems?
• How can they be identified or even predicted at an early stage?
• What risks do they involve?
• How must humankind act in order to prevent negative developments at the global level, to avert threats and/or mitigate the consequences of global change?

Global change research must therefore deal with the diagnosis, prediction and assessment of global trends, the prevention of negative trends, "repairing" existing damage (*remediation* and *restoration*) and adaptation to the unavoidable. Therefore, the primary interactions between these trends must be identified, described and explained.

Such research should be guided by the principle of sustainable development. The crucial element of this concept, now generally acknowledged, is the interdependence of environment *and* development (AGENDA 21). This reflects a growing insight that human beings and their environment are closely integrated within a system of mutual interaction. Research on global change is therefore confronted with two fundamental problems. Firstly, the investigation of the Earth System requires an *integrative* approach because the interactions between its components operate across the boundaries of single disciplines, sectors or environmental media. The second fundamental problem is the enormous *complexity* of the dynamic interrelationships involved, which makes a distinct description, any overall analysis and modeling much more difficult. The only approach capable of responding adequately to these problems is one that is networked and interdisciplinary. The sectoral approach must be supplemented by a *systems approach* that establishes linkages between different strands of research.

2.1.1
The Global Network of Interrelations

The Council has proposed a new *method* for holistic analysis of the present crisis of the Earth System (WBGU, 1993 and 1995a). The elements chosen for that analysis are not, as is often the case elsewhere, a set of easily indexed base variables, such as atmospheric concentrations of CO_2, population size or gross national product. Instead, the most important global trends are used as qualitative elements (*Fig. 5*). They are termed *trends of global change* and provide information about the dominant features of global development.

The development of the Earth System is then described using this set of trends. Trends refer to highly complex natural or anthropogenic processes but do not resolve the internal processes in detail. A precise analysis of micromechanisms is not necessary at the highly aggregated level of global change since these mechanisms have either no effect at all or only indirect impact on global changes in people-environment relations. Because the trends were formulated in such a way that they "overlap" as little as possible in their significance, it is possible to use them as basic elements of a systems analysis of the dynamics of global change.

Another prerequisite is the definition of indicators for the respective trends based either directly or

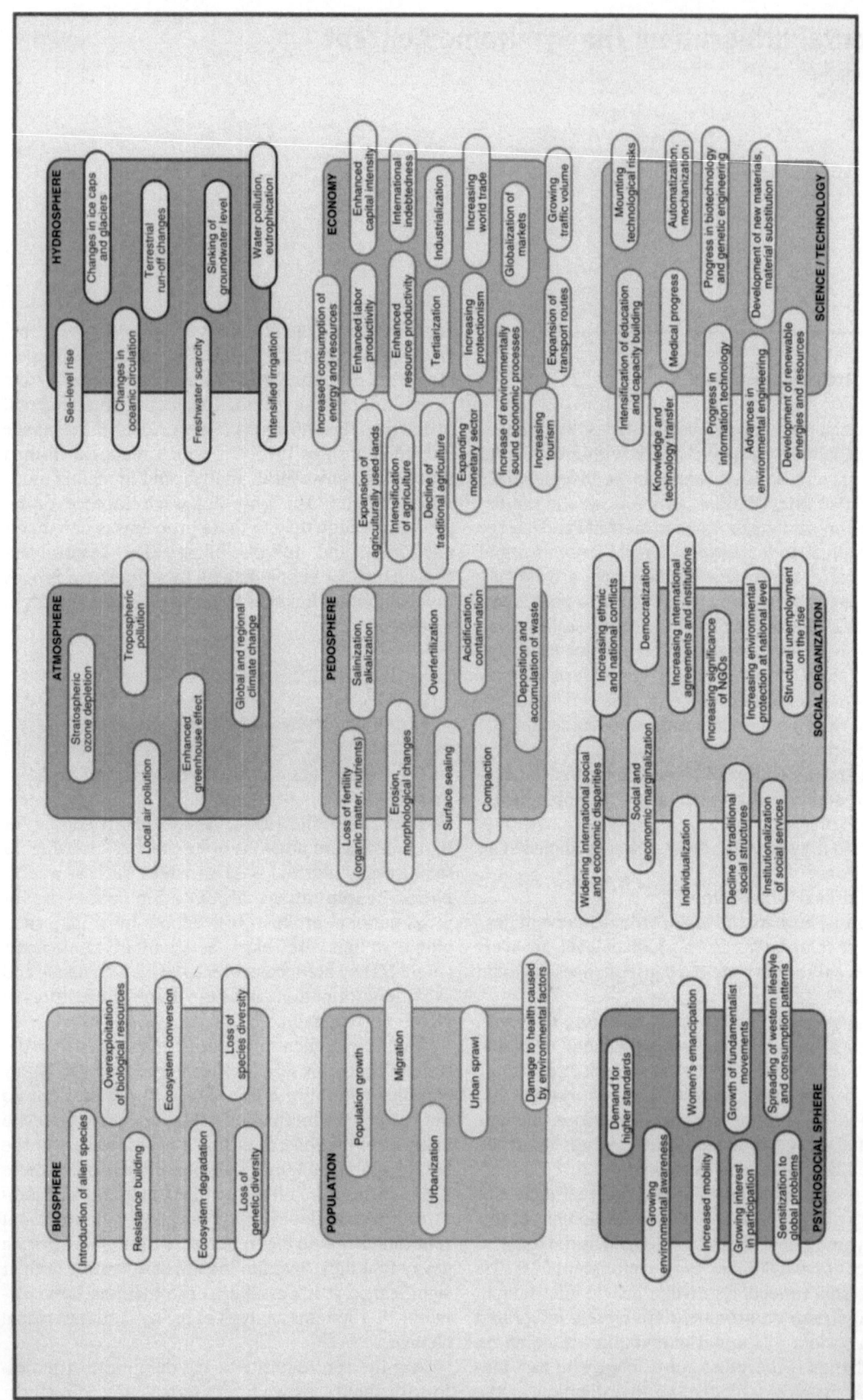

Figure 5
The Global Network of Interrelations.
Source: WBGU

indirectly on measurements (*Box 14*). These variables may be physical, chemical or biological parameters gained by observation, or they can be obtained through social scientific analysis. The important thing here is that available information need not be comprehensive – only references to the qualitative features are necessary.

Those trends possessing special relevance for global change are selected on the basis of educated guesses. They are not evaluated at the outset; problematic processes such as climate change, loss of species diversity or soil erosion are placed alongside other trends – like globalization of markets, or progress in biotechnology and genetic engineering – which can have positive or negative impacts depending on the perspective taken and the specific manifestation of the trend in question. Another category of trends are those which may lead to the mitigation of global problems, e.g. increasing environmental protection at

national level, growing environmental awareness, or the increase in international agreements.

Generally speaking, the focal points of this approach are also major topics in the public and international debate on global change. Some of these key topics, or *core problems of global change (Box 15)*, correspond exactly to specific trends, while others may be identified, in the form of "megatrends", as the sum of interrelated global tendencies. The core problem "soil degradation", for instance, is composed of several trends in the pedosphere (erosion, loss of fertility, salinization, surface sealing, etc.), while climate change represents a particularly dominant single trend in the network of interrelations.

It is not possible to assess trends or core problems in isolation from their cause-effect relations. The significance of core problems can only be ascertained by analyzing the overall context. To generate this context, the 80 or so trends of global environment

---

BOX 14

Environmental Indicators – Definitions and Applications

Environmental indicators are parameters for environmental perception and assessment. A large number of possible observations and a great deal of information must be systematized and *condensed* into key characteristics in order to be able to determine and evaluate the *current state* or the *trends* of the system considered. If the indicators are chosen correctly, even a fraction of the available data is sufficient to characterize or clarify a complex situation.

Examples:

- The "culprits" responsible for the reduction of stratospheric ozone concentrations can be identified directly on the basis of the concentration of reactive halogenide compounds in the stratosphere: these compounds are not formed in nature, but are decomposition products of CFCs.
- Tropical storms can only form over the near-Equator ocean if the surface water temperature exceeds a minimum temperature of 27°C.

However, individual measured variables or indicators are not enough to adequately describe or assess the state of the complex environmental system. For this reason, several types of indicators

are used, ranging in hierarchy from "simple" to "composite" to "systemic".

- *Simple indicators* are measured variables used, for instance, for substances with a high risk potential and for which synergistic or antagonistic properties have not yet been taken into consideration.
  Example: Dioxin concentration in exhaust gases.

- *Composite indicators* are specific combinations of system variables that have independent significance (e.g. aggregations of related or complementary characteristics) and which are able to indicate more complex properties of the particular system.
  Examples: When assessing *soil contamination*, it is possible to use the concentrations of potential contaminant groups (heavy metals, organic substances, radioactive substances, etc.) as indicators of the degree of pollution. The tendential development of a forest suffering from the new type of forest damage can be assessed by a group of empirical characteristics (thinning of crowns, discoloration of needles, excessive fruit growth, etc.).

- *Systemic indicators* provide information on relations and interactions which exist between simple and composite characteristics and which are derived non-additively from measured and observed variables. Special indicators include system characteristics such as complex-

▶

ity, stability, capacity to repair damage, development potential, degree of networking, feedback density, etc.

Example: The species diversity of an ecosystem (a tropical rainforest, for example) is an indicator of the degree of networking of this system; loss of species is thus an indicator of the threat to which the total ecological system is exposed.

Another dimension is obtained by distinguishing between "analytical" and "normative" indicators.

- *Analytical indicators* are key characteristics used to describe the state of the system under analysis. These variables can be plotted on a scale or spectrum through direct observation or measurement.
  Examples: Cadmium concentrations in samples taken from hazardous waste sites; occurrence of nettles as indicator plants for high concentrations of nitrogen.
- *Normative indicators* are required when the system is assessed in terms of external (ethical, political or economic) standards. It is then necessary to determine key characteristics that indicate the *quality* of a system's state or the *correctness* of a systemic trend. Normative indicators may similarly possess all degrees of complexity from simple to systemic.
  Example: Classification of a motor vehicle as "low-emission".

Analytical indicators can often be translated directly into normative indicators by specifying social preferences or goals. For example, the concentration of nitrate in groundwater acquires the function of a simple and normative key characteristic if it is based on a limit "x" stipulated by a government agency. Saying "smaller or larger than x" is then equivalent to "permissible or impermissible". "Respect for human rights" is a much more complex normative indicator, by contrast, and cannot be extrapolated directly from analytical characteristics.

In order to analyze the interactions between the ecosphere and the anthroposphere, it is necessary to devise additional indicators, complementary to those for the ecosphere, which measure economic, political and psychosocial conditions and trends (*social monitoring*). Much work is being done to develop such indicator systems, and a number of proposals have already been made for the economic sector specifically (e.g. a new type of overall environmental and economic assessment, estimation of the costs of neglected environmental protection). In addition to problems of definition, methodological questions relating to continuous, large-scale or random measurement and assessment of the respective characteristics must be clarified.

Environmental information systems are composed of indicators for all hierarchical levels. They form a well-defined framework that essentially serves to structure and organize information. The aim is to facilitate both the actual process of developing indicators as well as access to these indicators.

Generally speaking, any information system for Environment and Development (E&D) must meet the three basic requirements which follow:
- improvement of E&D information
- intensification of E&D communication
- support and verification of E&D policy.

Thus, a list of indicators primarily serves as an aid for objectivized perception and assessment of the environment. A whole range of different E&D information systems now exists worldwide, whereby the annual reports of the World Bank, the World Resources Institute, the Organization for Economic Cooperation and Development (OECD) and UNEP are of special importance for monitoring global change. In Germany, one of the main sources for information of this type is the Environmental Data publication of the Federal Environment Agency.

Because "sustainable development" is so difficult to define with any precision, expectations regarding the indicators to be applied are often unclear or exaggerated (*see Box 17*). Special attention must be given here to explicit definition of a reference frame, particularly where value judgments are concerned. This reference frame must integrate not only descriptions of the essential elements of the environmental system and its dynamics (variables, capacities and feedback loops) but also externally defined normative elements (environmental quality targets and guiding principles).

and development that have been documented thus far by the Council are linked to one another on the basis of expert knowledge by *determining interactions*. Every influence exerted by one trend on another is described in qualitative terms as "reinforcement" or "attenuation". It can be assumed, for example, that the anthropogenic enhanced greenhouse effect reinforces sea-level rise, or that the trend towards women's emancipation weakens population growth.

The various trends and their interactions can thus be combined in a qualitative *global network of interrelations*, which describes global change as a system and which represents the starting point for more extensive analysis of the Earth System's dy-

namics. With the help of this empirical-phenomenological description of global change, it is possible to produce qualitative models, this being the subject of a current BMBF research project.

2.1.2
Syndromes as Functional Patterns of Global Change

Networks of interrelations can be developed for other levels besides the global. A regionalized analysis of the Earth System using this instrument provides clear indication that the interactions in certain regions between human societies and the environ-

---

BOX 15

Core Problems of Global Change

ECOSPHERE

- *Climate change:* By enriching the atmosphere with long-lived greenhouse gases, humankind is inducing a significant level of climate change that can already be distinguished from natural climate variability "noise". There is growing anxiety that anthropogenic global warming is having feedbacks on oceanic circulation and the dynamics of the polar ice caps. Extensive uncertainty still prevails as to the precise impacts that the predicted shift of the climate belts (and thus vegetation cover and cultivation zones), rising sea level and increasingly frequent weather extremes will have on human societies and nature, both regional and globally.

- *Soil degradation:* In many countries today the soils of the Earth display degradation ranging from medium to extreme severity, and the situation is worsening from year to year. Such degradation is caused by rapid growth of the world population and its economic activities, resulting in overexploitation and transformation of plant cover, compaction and surface sealing of soils, as well as contamination by organic and inorganic compounds. Severe soil degradation means destruction of humanity's life-support systems and can therefore trigger famine, migration and military conflicts.

- *Loss of biodiversity:* Land-use changes spanning large areas of the globe (such as clearing

of forests, conversion of pastureland to cultivated land, etc.) bring about a reduction in the reservoir of potentially useful species and the natural products they provide, an impairment of the regulatory function of ecosystems and a decline in culturally and esthetically valuable biotopes. Loss of plant varieties and domestic animal breeds leads to greater susceptibility to pests and diseases, thus endangering the very food sources on which humanity is vitally dependent.

- *Scarcity and pollution of freshwater resources:* Freshwater resources are being overexploited on a local and regional scale through irrigation farming, industrialization and urban growth. Many parts of the world face mounting scarcity and pollution of water supplies. The consequence is a rise in economic, social and political conflicts over declining water resources, which in turn may have global impacts.

- *Overexploitation and pollution of the world ocean:* The oceans perform important ecological (especially climatic) functions, are a major source of food and act as a sink for anthropogenic wastes. Coastal regions and marginal seas, in particular, are further polluted with contaminants through immissions and direct discharges via rivers. Global impacts ensue, beyond the threats to fishing regions, due to the importance of fisheries for global food security.

- *Increasing incidence of human-induced natural disasters:* There are many indications that natural disasters are increasing in frequency as a result of human interference with natural systems. Forest clearing in the Himalayas, for

▶

example, gives rise to floods in foothill regions, thus posing an existential threat to the population there. Among other things, this induces migration pressure (environmental refugees) and the concomitant impacts on large sections of the international community.

ANTHROPOSPHERE

- *Population growth and distribution:* The world population continues to grow, primarily in the developing and newly industrializing countries. One of the root causes is inadequate education, which is bound up with high birth rates, weak social security systems and social marginalization of large parts of the population in these countries. Other trends are rural-urban migration and intra- and international migration flows. The latter produce rapid urban growth, particularly in coastal regions; the urban infrastructure (energy, water, transport, social services, etc.) of many cities is unable to keep pace with this growth. The environmental degradation and poverty which then result, and the potential for social unrest this entails, are having global impacts.

- *Environmental threats to global food security:* Large sections of humanity suffer from malnutrition and undernourishment. Feeding these people is rendered increasingly difficult by soil degradation, water scarcity and population growth. This trend is frequently reinforced by misdirected economic and development policies.

- *Environmental threats to health:* Factors such as population growth, famine, war, contamination of drinking water and inadequate waste water treatment lead to an increasing incidence of infectious diseases and epidemics in many countries of the world. As global mobility grows, so, too, does the risk of rapid spreading epidemics. In industrialized countries, air pollution causes increased incidence and severity of certain illnesses among the population.

- *Global disparities in development:* The structural imbalances between industrialized and developing countries have not declined in recent decades – on the contrary. The driving forces behind this development are economic, technical and social changes, above all the globalization of the world economy and the intensifying international division of labor. This process has helped some countries to achieve the desired economic development, though often at the expense of the natural environment. Nevertheless, most developing countries (particularly in Africa) have remained very poor, and it is there that the loss of social security and related migration processes are creating enormous problems. This "development dilemma" is characteristic of global change and represents a growing risk.

---

ment frequently operate along typical lines. These functional patterns (*syndromes*) are unfavorable and characteristic constellations of natural and civilizational trends and their respective interactions, and can be identified in many regions of the world. The Council's underlying thesis is that complex global environmental and development problems can be attributed to a discrete number of *environmental degradation patterns*.

Syndromes are transsectoral in nature: specific problems may affect several sectors (such as the economy, the biosphere, population) or environmental media (soil, water, air). They are always related, directly or indirectly, to natural resources. Syndromes are globally relevant when they modify the Earth System and have a noticeable impact, directly or indirectly, on the basis of life for a major part of humankind, or when global solutions are needed to surmount the problems.

Each one of these "clinical profiles" of the Earth System therefore represents a distinct basic pattern of environmental degradation induced by human society. This means – in theory – that the respective syndrome manifests itself independently of the others and can continue to unfold. This is particularly the case where syndromes exhibit self-reinforcement mechanisms, examples being the *Rural Exodus* and *Mass Tourism Syndromes*. If, in the former case, there is a general deterioration of rural infrastructure and the living conditions of the rural population as a result of outmigration, the pressure to migrate to cities is reinforced. In the latter case, the manifestation of the syndrome in a specific region renders it unattractive for tourists, who then look for new regions and attractions, thus spreading the typical pattern of destruction.

However, the basic autonomy of the syndromes by no means rules out the possibility of passive overlapping or active interaction between such degradation

A Typology of Syndrome Coupling

COINCIDENCE

The weakest, yet most frequent form in which syndromes interact is when they occur simultaneously in a country or region, but without one acting as a driving force for the other. A country like Australia, for example, may be affected by the *Katanga Syndrome*, the *Dust Bowl Syndrome* and the *Mass Tourism Syndrome* at the same time, without any significant mutual reinforcement occurring between them. Nevertheless, this coincidence of syndromes can still have profound importance, in that *hot spots* of global change can be identified in this way. Moreover, such "weak" links can be important when assessing the general vulnerability of a country to global change. In countries low in "defenses" (natural resources, capital, know-how, stable political situation, etc.), the coincidence of only two syndromes is sometimes enough to completely overtax its "resistance" and spontaneously trigger additional syndromes (e.g. the *Scorched Earth Syndrome*).

COUPLING THROUGH COMMON TRENDS

A stronger form of syndrome linkage is when two syndromes have one or several common key trends. If, as in the case of the *Sahel* and the *Rural Exodus Syndrome*, the trend of "social and economic marginalization" is a component part of the core mechanism, the parallel occurrence of the two syndromes in space and time will be regarded as more than pure coincidence – especially if the global trend of marginalization is explained by them to a large extent.

INFECTION

A syndrome already active may trigger another syndrome in a certain region. Deliberate reshaping of the natural environment through large-scale projects (*Aral Sea Syndrome*), for instance, may lead to changes in people-environment interactions in the region concerned and cause the *Rural Exodus Syndrome* and/or the *Sahel Syndrome* to emerge, even though these degradation patterns did not exist there previously.

REINFORCEMENT

Trends can have a reinforcing (or attenuating) effect on each other, but so can entire syndromes. In this case they do not trigger off other syndromes through common trends but through the total force of their characteristic pattern. An example of this is the driving force exerted by the *Sahel Syndrome* on the *Favela Syndrome*. The simultaneous incidence of phenomena such as soil erosion, marginalization of the rural population and growth of urban agglomerations which can be observed in newly-industrializing and developing countries, in particular, is not a mere spatial coincidence, but reflects a syndrome-reinforcing linkage of immense global relevance. Another example is the reinforcing effect exerted by expanding urban and infrastructural centers (*Urban Sprawl Syndrome*) on the disposal of waste (*Waste Dumping Syndrome*).

ATTENUATION

Syndromes may also be linked through attenuation. An example of this is the impact that the *Scorched Earth Syndrome* has on the *Mass Tourism Syndrome*; whenever wars and civil wars involve deliberate destruction of civilizational infrastructures and the natural environment, recreational tourism depending on the latter declines immediately as a result. Former Yugoslavia is the most recent example of this phenomenon. The converse example is the "death strip" along the former intra-German border, where nature was able to develop relatively unhindered for many years, thus escaping the potential damage caused by the *Urban Sprawl Syndrome* or the *Dust Bowl Syndrome*, for instance.

SUCCESSION

Syndromes are, of course, part of the historical development of the *people-environment interface*. If the syndrome approach is used to analyze retrospectively the history of human use of and damage to nature – and the history of the environment in recent years provides sufficient material for this purpose – it is possible to identify not only past occurrences of individual syndromes (the ironworks of Saxony induced the *Smokestack Syndrome* in the early 19th century, for example), but also typical *succession patterns* of syndromes. The sequence of development stages through which human civilization progresses is evidently linked to very specific syndromes, which can thus be used, at least in an exploratory way, to assess the future development of the Earth System. One such succession of syndromes commences with the *Sahel Syndrome*, branches off to the *Green Revolution Syndrome* or the *Asian Tiger Syndrome* and later from the *Dust Bowl* or *Urban Sprawl Syndrome* to the *Waste Dump Syndrome*.

patterns. One can distinguish between several forms of syndrome coupling (*Box 16*).

The syndromes, or "clinical profiles" can be plotted as maps, which then show where and how strong the respective syndrome is. If, for example, one gives each individual syndrome a different color with several shades of intensity, then superimposing the relevant maps should provide a meaningful picture of the state of environment and development on Planet Earth. A specific example in this context is the *Sahel Syndrome*, which is described in detail in *Section C 6* below.

The syndrome concept offers several options: firstly, analysis can be pursued to such an extent that the vulnerability of a given region to a given syndrome can be determined (*prevention*). Secondly, the systemic integration of causes, mechanisms and effects as a problem-specific pattern produces an improved understanding of the system as a whole, thus enabling sound recommendations for *curing* the syndromes.

Finally, the concept opens up a way to *operationalize* the notion of sustainable development, which generally refers to an acceptable coevolution of the ecosphere and the anthroposphere. To characterize global development, undesired or hazardous conditions in the environmental, economic, social and cultural spheres are defined. These "non-sustainable domains" are demarcated from the permissible action space by "crash barriers" (or "boundary surfaces" in

a 3-dimensional view). Within the latter, society remains capable of taking action and making free decisions regarding human activities. Only when society approaches the boundary surfaces is there a higher risk and diminished stability, whereas a drift of the Earth System beyond the crash barriers should be avoided at all cost (*Fig. 6*).

The complexity of the system and the uncertainty that is often inherent in available data mean that "crash barriers" and "boundary surfaces" cannot be defined with precision. They must therefore be regarded as "boundary zones" with fuzzy contours. Estimates of where these zones are located will vary according to the current state of knowledge, prevailing values and the willingness of the population to accept risks, so they will tend to shift over time. The job of Earth System management is to stop any drift into the space beyond the crash barriers.

The Council has applied this theoretical approach in its climate protection scenario to derive global targets for CO_2 emission reductions (WBGU, 1995b). The first step was to define the limits to global climate change so that general principles such as "preservation of Creation" are not violated. These limits take the form of a "tolerable window" for the global climate system that must be complied with if unacceptable consequences for humankind and the environment are to be avoided. Assuming that a departure from this climate window is prohibited, i.e. non-

Figure 6
The "crash barrier" within the syndrome concept. The "crash barriers", or boundary surfaces, demarcate the domain of free action from the "non-sustainable" domain in which the syndrome is manifest.
Source: WBGU

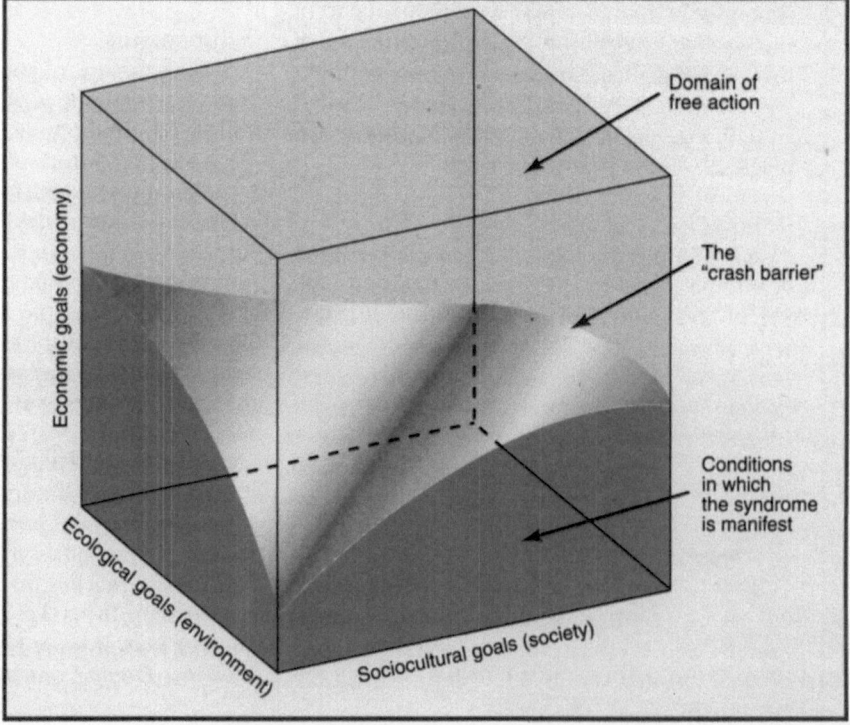

sustainable, it is possible to calculate the CO_2 emission reduction rates that are necessary in the future.

This example is a clear illustration that a problem-oriented systems approach can also facilitate the operationalization of the sustainable development concept. The syndrome concept provides an excellent basis for implementation of the crash barrier model – sustainable development can then be described as the absence or mitigation of syndromes. Syndromes at regional level can then be assessed in relation to this (Utopian) ideal case by determining their "distance" from the desired condition using systemic indicators (*Box 17*).

The syndrome concept provides a new basis for global change research, the knowledge base of which continues to be split up according to the environmental media or core problems. This sectoral or disciplinary approach is certainly justified: without searching for a deeper understanding of the individual problem areas and their functional mechanisms, it is impossible to understand the specific aspects of environmental stress.

As has been repeatedly mentioned in this Report, environmental research must also be carried out on a synthetic and integrative basis if it is to be system- and benefit-oriented. The global perspective, in particular, calls for joint work on the part of various disciplines, interest groups and actors. In the view of the Council, global change research must be based from the initial design stage onwards on such interdisciplinary structuring of problems and solutions.

The syndrome concept elaborated by the Advisory Council provides new, concrete options for shaping research activities. Given the desired conditions for global change research referred to in *Section B 3.10* – *interdisciplinarity*, *internationality* and *problem-solving competence* – the obvious choice is to structure future environmental research along transdisciplinary lines. It is therefore recommended that these syndromes be adopted as the central objects of future global change research.

2.2
List of Global Change Syndromes

Identifying and describing the main syndromes involving disturbances to or non-sustainable forms of people-environment relations is an essential prerequisite for the response options mentioned in *Section C 1.1* and for a new research orientation. The following typology is intended as a start in this direc-

---

BOX 17

Syndrome Profiles as Indicators of Sustainable Development

Before the syndrome concept can be operationalized, it is necessary to devise indicators which base the identification and assessment of global change syndromes on a standardizable method. This involves compiling lists of indicators that link the defining features of the respective syndrome (quantitative dimensions, symptoms, system characteristics, evaluation of damage) to one another and which make up a specific *syndrome profile*.

In terms of the classification scheme in *Box 14*, these syndrome profiles are *normative-systemic* indicators. They include composite and systemic indicators in addition to normative elements, because not only do they identify complex and unsustainable trends within the global environment and development process, they also evaluate them. These profiles could thus be put to immediate use as "*sustainability indicators*"; development is characterized as sustainable when the global or regional manifestation of individual or all syndrome profiles remains within certain limits. Work must be carried out in the following steps:

- Allocation of responsibility for the various core problems (soil degradation, greenhouse effect, etc.) to the causal syndrome in proportion to type and amount.
- Grouping and detailed adjustment of existing indicator list (lists of hazardous substances, etc.) in line with the classification scheme for syndromes.
- Elaboration of key criteria for determining which syndrome is more severe than others (degree of harm to people and nature, curative potential, degree of impact, etc.).

Proceeding in this way enables a clear assessment to be made of the current intensity of a syndrome and its general "hazardousness". Armed with this information, it is possible to deploy available resources in a more targeted manner – firstly to mitigate those syndromes that obviously have disastrous consequences for humankind and nature, and even to prevent them altogether where possible (*crash barrier scenario*).

BOX 18

Overview of Global Change Syndromes

"UTILIZATION" SYNDROMES

1. Overcultivation of marginal land: *Sahel Syndrome*
2. Overexploitation of natural ecosystems: *Overexploitation Syndrome*
3. Environmental degradation through abandonment of traditional agricultural practices: *Rural Exodus Syndrome*
4. Non-sustainable agro-industrial use of soils and bodies of water: *Dust Bowl Syndrome*
5. Environmental degradation through depletion of non-renewable resources: *Katanga Syndrome*
6. Development and destruction of nature for recreational ends: *Mass Tourism Syndrome*
7. Environmental destruction through war and military action: *Scorched Earth Syndrome*

"DEVELOPMENT" SYNDROMES

8. Environmental damage of natural landscapes as a result of large-scale projects: *Aral Sea Syndrome*

9. Environmental degradation through the introduction of inappropriate farming methods: *Green Revolution Syndrome*
10. Disregard for environmental standards in the course of rapid economic growth: *Asian Tigers Syndrome*
11. Environmental degradation through uncontrolled urban growth: *Favela Syndrome*
12. Destruction of landscapes through planned expansion of urban infrastructures: *Urban Sprawl Syndrome*
13. Singular anthropogenic environmental disasters with long-term impacts: *Major Accident Syndrome*

"SINK" SYNDROMES

14. Environmental degradation through large-scale diffusion of long-lived substances: *Smokestack Syndrome*
15. Environmental degradation through controlled and uncontrolled disposal of waste: *Waste Dumping Syndrome*
16. Local contamination of environmental assets at industrial locations: *Contaminated Land Syndrome*

---

tion, and itself needs substantial further research (addressing, inter alia, the divisions and links between the various syndromes). The Council has already applied its syndrome concept to the problem of anthropogenic soil degradation (WBGU, 1995a). The list of syndromes presented here represents a further development of the 1994 version: the "clinical profiles" are no longer focused on soils alone, but encompass all compartments of the ecosphere and anthroposphere simultaneously.

We distinguish between three major groups of syndromes:

1. Syndromes resulting from inappropriate use of natural resources as production factors (*"Utilization" Syndromes*).
2. People-environment problems arising from non-sustainable development (*"Development" Syndromes*).
3. Environmental degradation through society's use of non-adapted disposal systems (*"Sink" Syndromes*).

Within these groups, various archetypal patterns of global environmental problems can be identified

(*Overview in Box 18*). All syndromes must meet the following criteria, however:

- Each syndrome relates directly or indirectly to the environment; exclusive reference to core problems within the anthroposphere is not permitted.
- The syndrome should occur as a visible or virulent cross-cutting problem in many regions of the world.
- The syndrome should describe non-sustainable development and/or significant environmental degradation.

2.2.1
"Utilization" Syndromes

OVERCULTIVATION OF MARGINAL LAND: SAHEL SYNDROME

The *Sahel Syndrome* involves a complex web of factors causing environmental degradation when the ecological carrying capacity is exceeded in regions where natural environmental conditions (climate, soil) restrict agricultural use (marginal locations)

(WBGU, 1995a). Typical manifestations of this pattern are soil degradation (erosion, loss of fertility, salinization), the spread of desert-like conditions (desertification), the depletion of fossil aquifers, the conversion of semi-natural ecosystems (due to deforestation, for example), the loss of biodiversity and changes in regional climate.

The *Sahel Syndrome* typically appears in subsistence economies where groups of rural poor and sections of the population threatened with marginalization are confronted with increasing degradation of their natural environment due to overexploitation of agricultural land (e.g. overgrazing, spread of farming to ecologically sensitive regions). The syndrome-specific problems of the population include mounting poverty, rural exodus, greater vulnerability to food crises as well as rising frequency of political and social conflicts over scarce resources. The replacement of sustainable farming with intensified land management methods, such as abandonment of crop rotation systems or shortened fallow periods, is an important element of the syndrome. Unwise development strategies (sedentarization of nomads, construction of deep wells) may also operate as contributory factors. Development of the syndrome is reinforced by rapid population growth, and occurs within the wider context of social transformation, as evidenced by the collapse of traditional solidarity systems, disruption of local price mechanisms by subsidized exports from industrialized countries, and cultural transformation. In the course of the *Sahel Syndrome*, the scope for action on the part of the affected social groups gradually narrows (to severe famine in extreme cases) due to the mutual reinforcement of poverty, overexploitation and environmental degradation.

In the Sahel zone itself, more than half the population is threatened by starvation following destabilization of rural production and social systems. As a result of population growth, traditional crop rotation methods have approached their critical limits, forcing an expansion of agricultural production to marginal lands. The consequence of such inappropriate land use is desertification and rural-urban migration.

Another typical trend within the *Sahel Syndrome* is forest conversion at marginal locations and subsequent exploitation by subsistence farming – otherwise known as *shifting cultivation* or *slash-and-burn agriculture*. In southern Thailand, for example, severe floods caused by soil erosion are a direct consequence of this form of land use in the northern part of the country.

Symptoms: Destabilization of ecosystems, loss of biodiversity, soil degradation, desertification, threats to food security, marginalization, rural exodus.

OVEREXPLOITATION OF NATURAL ECOSYSTEMS: OVEREXPLOITATION SYNDROME

The *Overexploitation Syndrome* involves the conversion of natural ecosystems and the overexploitation of biological resources, and affects both terrestrial (forests, savannas) and marine (overfishing) ecosystems. The common feature is that ecosystems are overexploited without regard for their regenerative capacity, resulting in severe damage to the natural balance. Violation of the sustainability principle leads to degradation and even destruction of natural ecosystems, e.g. through outright clearance of forests, overgrazing of otherwise uncultivated land, or overfishing. The immediate consequences are loss of habitat, the resultant loss of biological diversity, and erosion (particularly in mountainous regions). The latter causes a substantial increase in susceptibility to natural disasters (landslides, floods), and in the amount of sediment transported by rivers, thus leading to floods and other threats to coastal ecosystems, not to mention the enormous costs incurred through the silting up of navigable channels and harbors. In addition, the release of CO_2 from biomass and soils enhances the greenhouse effect. For the local population, the conversion of ecosystems means loss of livelihood, resulting in impoverishment and loss of cultural identity. One of the typical features of the *Overexploitation Syndrome* is the permitting of overexploitation for short-term gains. What is more, this type of land management (often involving multinational corporations) usually leads to profits being transferred to big cities or out of the country. Local economies incur substantial costs, but earn minimal profits in return. These mechanisms and their negative impacts have now been recognized, and various responses aimed at counteracting them have been initiated by NGOs and international agencies (FAO, Biodiversity Convention).

Typical manifestations of the *Overexploitation Syndrome* include clear felling of tropical rainforests and subsequent land-use changes (in Brazil, Malaysia, Indonesia, Myanmar, etc.), or the clearing of mangroves in the tidal zone of tropical coasts. Another severe case is the overexploitation of boreal forests having a low regenerative capacity. Siberia, for instance, permits the large-scale clearance and destruction of boreal coniferous forests using modern harvesting equipment. Similar mechanisms lead to overfishing of the world's oceans. With the help of technologically advanced but ecologically disastrous catching methods, all 17 major fishing grounds are being fished at or beyond the limit of their capacity; stocks have been severely decimated in 13 of them.

Symptoms: biodiversity loss, climate change, freshwater scarcity, soil erosion, increasing incidence of natural disasters, threat to food security.

ENVIRONMENTAL DEGRADATION THROUGH ABANDONMENT OF TRADITIONAL AGRICULTURAL PRACTICES: RURAL EXODUS SYNDROME

The *Rural Exodus Syndrome* refers to environmental degradation caused by the abandonment of previously sustainable land-use practices. Labor-intensive methods for cultivating small plots of land, such as terraced slopes, small-scale irrigation systems, or protective measures against wind erosion, become increasingly unprofitable as socioeconomic conditions change. The reason is often the exodus of young males to urban centers in search of higher wages, better educational opportunities and a less "provincial" way of life (*see Favela Syndrome, Asian Tigers Syndrome*). Tilling the land is too labor-intensive for the women, children and elderly persons who are left behind. Extensification (expanding the area of land under cultivation), and the associated lack of protective measures and proper care of the land, causes erosion (often reinforced by excessive timber-felling on steep slopes, *see Sahel Syndrome*), mud-rock flows and landslides. The ultimate result is loss of arable land, and disruptions or destruction of supply and communication networks. Wind erosion may increase drastically in other regions if protective action is not taken. The *Rural Exodus Syndrome* jeopardizes the livelihood of subsistence farmers. At the same time, there is a growing dependence on external transfers of goods, and remittances from migrants.

This process can be observed in its ideal-typical form in the Karakoram Mountains in northern Pakistan. The construction of access roads to this remote region induced rapid growth in the external flows of goods and commodities. Improved education for children, combined with seasonal and permanent labor migration of men, particularly to Karachi, led to neglect and decline of what used to be intensive land management practices. Modernization of agriculture (partial mechanization, e.g. of harvest work) could not compensate fully for labor force shrinkages, so the outcome was decline of both cultivated areas and productivity, thus endangering the subsistence of food producers. In extreme cases, entire village communities abandoned their land following landslides and increasingly frequent avalanches. Similar problems occur with labor-intensive wet rice terraces on steep slopes (e.g. in northern Luzon, Philippines) and on the fertile slopes of Mt. Kilimanjaro.

Symptoms: loss of biodiversity, soil erosion, rural exodus, threat to food security, marginalization.

NON-SUSTAINABLE AGRO-INDUSTRIAL USE OF SOILS AND BODIES OF WATER: DUST BOWL SYNDROME

The *Dust Bowl Syndrome* is a specific causal complex in which environmental destruction is caused by non-sustainable use of soils or bodies of water as biomass production factors, involving intensive deployment of energy, capital and technology.

Modern agriculture seeks to achieve the greatest possible yields on the areas available to it. The long-term repercussions for the environment are typically ignored in the pursuit of short-term goals. In a broader sense, the *Dust Bowl Syndrome* is also characteristic of some types of forest management (e.g. planting and then clearing of rapidly growing monocultures without regard for soil degradation or biodiversity loss) and aquaculture (eutrophication, destruction of coastal ecosystems) when these are driven by similar motivations.

High-yielding varieties, agrochemicals and mechanization form the basis of modern industrial biomass production. Farming enterprises in such agricultural systems are highly mechanized and automated (e.g. intensive livestock farming, modern irrigation systems, aquaculture, forest monocultures) and require only a small workforce. Commercial success depends on the right combination of capital, know-how, political support (e.g. land consolidation or other large-scale projects, see *Aral Sea Syndrome*) and favorable location factors. The central mechanism driving the syndrome is technological and innovative competition over regional and, increasingly, global markets for agricultural produce. This competition is greatly distorted by trade barriers and inadequate internalization of environmental effects. The syndrome is exacerbated further by high subsidies for energy, raw materials and manufacturing supplies (e.g. EU, North America).

Non-sustainable production methods can severely damage the environment, but not all can be assigned to the Dust Bowl Syndrome, since non-sustainable management of aquacultures is another component part of the latter. The type of damage ranges from altered hydrological conditions, eutrophication and contamination of surface and ground water to loss of biodiversity (loss of habitat, reduced networking of semi-natural ecosystems, decline in species diversity, isolation of wild fauna and flora, genetic erosion, etc.) and increased pesticide concentrations in the food chain, with all the damage to health and greenhouse gas emissions (CO_2, methane) this involves.

Symptoms: loss of ecosystem and species diversity, genetic erosion, eutrophication, acid rain, greenhouse effect, contamination of water bodies and air, freshwater scarcity, soil degradation, marginalization, rural exodus.

Environmental Degradation Through Depletion of Non-Renewable Resources: Katanga Syndrome

The *Katanga Syndrome* stands for the environmental damage caused by intensive mining of non-renewable resources above and below ground, with no consideration given to preservation of the natural environment.

Although mining of non-renewable resources is usually effected over limited periods (decades), it leaves behind permanent, sometimes irreversible environmental damage in many cases. A distinction can be made between two manifestations of the syndrome, namely the environmental impacts resulting from toxicity (release of small amounts of highly toxic substances, such as mercury), and the morphological and energy-related impacts that result when extreme masses of material are moved in order to extract either very large volumes of raw materials (gravel, brown coal) or very valuable but highly dispersed resources (such as blind rock in the case of diamonds or precious metals).

A typical feature of the syndrome is large-scale destruction of natural ecosystems and arable soils, particularly in the case of open-cast mining in developing and newly industrializing countries. Interim storage of these soils is prescribed by law in almost all industrialized countries, by contrast. Other effects include changes in morphology and subsidence of the land surface. This, in turn, has severe impacts on hydrological processes, such as surface runoff, increased sediment pollution in rivers and the groundwater table, as well as on soil erosion. The release of toxic substances leads to contamination of soils, surface and ground water, and the concomitant effects this has on biodiversity. The negative consequences for the local population range from serious damage to health to forced migration or resettlement, as is occurring with indigenous peoples in the Amazon "gold rush region". In general, the Katanga Syndrome is especially intense wherever, for want of investment capital, mining operations involve obsolete technologies low in energy and raw material efficiency.

Such non-sustainable use of natural resources is widespread. Examples in Germany are those regions where open-cast mining is carried out on a large scale; major "hot spots" of ore mining include Irian Jaya in Indonesia, Carajás in Brazil (iron ore, aluminum), Bougainville in Papua-New Guinea (copper) and, of course, Katanga. Oil prospecting and production (Nigeria, the Gulf states, Russia) also involve significant risks to the environment (oil spills, burning off natural gas, soil contamination due to leaks in pipelines).

Symptoms: loss of biodiversity, local air pollution, freshwater scarcity, change in runoff, water pollution, soil degradation, creation of contaminated sites, negative effects on health due to pollution.

Development and Destruction of Nature for Recreational Ends: Mass Tourism Syndrome

The *Mass Tourism Syndrome* describes the network of causes and effects generated by the steady growth of global tourism in recent decades and which leads to major environmental degradation in certain regions of the world. Typical "hot spots" are coastal areas and mountainous regions. Winter sports and pony trekking, for example, cause destruction or impairment of plant cover and tree vegetation, leading to biodiversity loss and soil erosion when reinforced by mechanization and other types of interference with the balance of nature (leveling, modifications to terrain, snow cannons), and hence a greater danger of landslides and avalanches. Mass tourism involves, for example, the conversion of semi-natural areas through the construction of touristic infrastructure (hotels, holiday homes, transport routes) and damage to or loss of sensitive mountain and coastal ecosystems (e.g. dune landscapes, salt-water marshes). The rapid growth of long-distance air travel in recent years causes pollution of the Earth's atmosphere. In the regions affected – especially on islands – the demand for freshwater is greatly increased (swimming pools, high levels of water consumption by tourists). Typical impacts include overexploitation of freshwater resources, which raises the specter of the regions' livelihood being destroyed through exhaustion of groundwater stocks, dessication of soils and erosion. The substantial and often seasonally varying stress on tourist regions results in serious problems regarding sewage treatment and disposal, with contamination and eutrophication of surface water or coastal ecosystems the possible consequence. Waste disposal problems are also on the rise.

The growing volume of tourism is directly induced by rising incomes in the industrialized nations and falling transport costs, combined with a simultaneous reduction in working hours and fundamental changes in leisure behavior. Another important factor is the increasing ease with which virtually any place in the world can be reached, not only as a result of infrastructural development, but also because the distances involved are no longer seen as prohibitive. In addition, a broad complex of psychological factors can be identified as playing a role (greater need for recreation due to mounting noise and other forms of pollution in urban areas, long-distance travel as status symbol, higher levels of education and thus greater interest in foreign cultures, discovering new places in the world, etc.).

People thus destroy the very thing they are looking for, namely unspoilt nature. These selfsame effects of tourism lead many to seek out new, unspoilt places to spend their holidays, thus spreading the syndrome to other regions at an accelerating pace.

Typical examples are the overdevelopment of previously semi-natural areas in Spain (Costa del Sol, Lanzarote), and the consequences of trekking tourism in Nepal.

Symptoms: loss of biodiversity, enhancement of the greenhouse effect by air travel, lack of freshwater supply, soil erosion, inadequate disposal of sewage and waste, fragmentation of landscapes by settlements, high consumption of resources.

ENVIRONMENTAL DESTRUCTION THROUGH WAR
AND MILITARY ACTION: SCORCHED EARTH
SYNDROME

The environmental degradation resulting from the direct and indirect impacts of military activities exhibits certain unique characteristics. The effects of maneuvers, regionally confined military operations and contaminated military sites constitute an additional problem within global change. The growing potential for regional conflicts and the hegemonial claims of global players lead to more hot spots where the local environment may incur permanent damage. A distinction can be made between the following "sub-syndromes".

Regional conflicts conducted with less advanced military equipment often produce almost irreversible forms of environmental degradation due to the increasing deployment of land mines. Anti-personnel weapons of this kind can be obtained at a very low unit price, are difficult to defuse and therefore constitute a serious long-term threat.

Another problem area is when advanced military machinery is used to intervene in local conflicts or to implement resolutions made by the international community. The weaker party in terms of military technology can hold local environmental resources to ransom, as the Iraqi army demonstrated with the oil fields in Kuwait. Other forms of blackmail using advanced technologies are also conceivable, such as (nuclear) power stations, dams, etc., or the threat to contaminate soil and water. A "scorched earth policy" may therefore have disastrous consequences for humanity and nature at local level, and may indeed have severe global impacts.

A third sub-syndrome exists in many parts of the world as a direct consequence of the arms race previously conducted by the two superpowers in East and West. Contaminated military sites now dot the landscape along the former boundaries. In the West, local environmental resources (soil, groundwater) will be exposed to these latent threats for many years to come, despite huge investments and modern technology. In the former Warsaw Pact states, these threats are much greater, given the lack of capital for decontamination work and the fact that the resultant disasters come to the attention of the world public much less.

Symptoms: loss of biodiversity due to chemical warfare agents (e.g. *agent orange*), permanent soil degradation due to mining, contamination caused by fuels and explosives, health hazards, greater flows of refugees.

2.2.2
"Development" Syndromes

ENVIRONMENTAL DAMAGE OF NATURAL
LANDSCAPES AS A RESULT OF LARGE-SCALE
PROJECTS: ARAL SEA SYNDROME

The *Aral Sea Syndrome* describes the failure of large-scale, extensive reshaping of semi-natural areas. Entire landscapes are affected by deliberate and systematically planned interventions involving major capital investments, but inadequate consideration of local conditions. The objective, as a rule, is the attainment of strategic targets defined in national or occasionally international policy frameworks, and implemented with the help of central planning in the form of large-scale projects. Many of these involve the construction of dams, irrigation systems and similar, but poor understanding of systemic interrelations means that impacts are given too little consideration, leading to environmental degradation and, in many cases, to severe disturbances of the social fabric. Other manifestations of the syndrome are when entire landscapes are adapted to mechanized agriculture through large-scale consolidation of arable land. A common feature of all these projects is the incapacity of planners to assess or manage the impacts of large-scale projects.

The Aral Sea, once the fourth largest freshwater lake in the world, manifests the syndrome in all its features. Fishing and agriculture were carried out in what was once a fertile region abundant in forests and species. For 30 years, however, the feeders of the Aral Sea have been tapped (only about 10% reaching the sea) and diverted to a gigantic irrigation system for cotton production. The surface area of the Aral Sea shrunk to half its former size, and the volume declined by two thirds. What used to be the bottom of the lake is now a salt desert, from which wind transports 40 to 150 million tons of salt and sand annually to the fertile land of the Amu Darya delta. All 24 species of fish are now extinct, meaning loss of employment for 60,000 fishermen. Short-term expansion of agricultural production has led to such severe environmental damage that extensive parts of the region have been desertified.

Other examples of the syndrome are large-scale dam projects (e.g. the Hoover, Assuan, Narmada and Bakun Dams, or the three-canyon project on the Yang Tse). Here, too, the social and environmental effects were either completely ignored or wrongly assessed.

An extreme example of such mega-projects was the Soviet plan to divert the large Siberian rivers to the south and in this way irrigate those parts of central Asia suffering from water scarcity. One of the main reasons for abandoning the project, which had already reached an advanced stage of planning, was the risk of drastic impacts on the world climate due to reduced discharge of water from the Arctic Ocean, and the resultant threat of the entire global circulation system collapsing.

Symptoms: loss of biodiversity, local or even global climate change, shortage of freshwater, soil degradation, forced resettlement of local population, danger of international conflicts, e.g. between riparians for water rights.

ENVIRONMENTAL DEGRADATION THROUGH THE INTRODUCTION OF INAPPROPRIATE FARMING METHODS: GREEN REVOLUTION SYNDROME

The Green Revolution Syndrome circumscribes the extensive, centrally planned modernization of agriculture with imported agricultural technology to secure an adequate food supply for a rapidly growing population, with negative impacts on the natural basis for production, on the one hand, and social structure, on the other. The Green Revolution, in which national development efforts run parallel to the activities of international donor organizations, helped many developing countries to increase significantly their agricultural yields. Typical Green Revolution technology involves simultaneous use of high-yielding cereal varieties, agrochemicals (commercial fertilizers and pesticides) and machines (tractors, harvesting machines, irrigation pumps, etc.). The three factors must be deployed in conjunction, however, thus necessitating substantial capital investments and agricultural consulting.

Supplying an exponentially growing population in the developing countries with sufficient food would never have been possible without the Green Revolution. The central mechanism is the "race" between population growth and the compulsion to increase food output by intensifying agricultural production. It soon transpired, however, that the Green Revolution was generating ecological and socioeconomic problems as well, due to the import of alien production methods and above all the incorrect application of the latter. In recent years, especially, dependence on imports, lack of foreign exchange and price increases have led to a situation in which the necessary complementarity of the three factors mentioned above is no longer ensured. When this is compounded by a lack of education among many farmers and inadequate consulting, the outcome is inappropriate use of techniques, leading in many cases to environmental degradation, e.g. through overfertilization or incorrect deployment of machinery and irrigation techniques. Very often, damage to health is caused by misuse of pesticides. A fundamental problem is rapid genetic erosion among cultivated plants when numerous indigenous varieties adapted to local conditions are displaced by high-yielding varieties requiring increased chemical protection. In addition, the Green Revolution reinforces regional economic disparities in that it usually only succeeds in traditional irrigation areas, but not in the arid zones of Asia and Africa.

A typical example of the Green Revolution is India, where it was introduced as a national program for rural development in the late 1960s. The program's successes were concentrated in areas where irrigation methods had been traditionally used, such as the large delta regions of the Ganges, Cauvery and Indus rivers. Enhanced yields, particularly in wheat farming, made India self-sufficient in food production, despite its rapidly increasing population. Now, however, there are visible signs of environmental damage causing a reduction of cropland.

Symptoms: loss of biodiversity, genetic erosion, groundwater pollution, soil degradation, threats to food security, health hazards through pesticide use, marginalization, rural exodus, reduction of cultural diversity, reinforcement of regional economic disparities.

DISREGARD FOR ENVIRONMENTAL STANDARDS IN THE COURSE OF RAPID ECONOMIC GROWTH: ASIAN TIGERS SYNDROME

Many regions in the so-called newly industrializing states are experiencing extremely rapid economic growth, or are in the process of accomplishing this structural transformation with a new kind of intensity and momentum that has grave implications for people and nature. The "Manchester Syndrome", which commenced in England in the mid-18th century and required more than 100 years to unfold, was the first example of the environmental problems generated by rapid development; the root problem in those days was lack of knowledge about environmental impacts and the absence of appropriate technologies.

The problem with the *Asian Tigers Syndrome,* in contrast, is that the development process has been drastically shortened, and therefore requires colossal efforts to contain the environmental degradation this involves. The mobility of the capital market, the globalization of markets, high transport capacities worldwide ·and locational advantages at local level, such as economic and political stability, low wages, lack of employee participation rights and initially lower standards of consumption, are all major factors which predispose a region to this syndrome. In addition, the constantly growing availability of production software (blueprints, production techniques, etc.) provides for economic growth of previously unknown proportions within increasingly shorter periods of time. It is patently obvious that neither the creation of adequate infrastructures for supply and disposal nor the introduction of suitable environmental technologies can keep pace with this growth. This applies above all to countries that try to imitate the Asian Tigers.

Although political failures are much to blame for these disparities in development, the latter are essentially the consequence of the explosive nature of such development and its internal dynamics. Investments would be necessary on a large scale to prevent irreparable ecological damage. Funds of this order are indeed available, but they are invested primarily in further economic growth – this being the only way to keep the process going. The consequences are visible in many affected regions in Southeast Asia, and are certain to emerge in India, Central and South America. Cities like Bangkok, Manila, Mexico City, Jakarta and Mumbai (Bombay) are already considered to have lost control of their traffic systems. Extreme local air pollution (smog), inadequate sewage treatment and environmentally harmful waste management are typical features of the syndrome, as is enormous consumption of resources and energy.

The Asian Development Bank estimates the cost of environmental damage up to the year 2000 at over US$ 2,500 billion, but this figure does not include other regions that will soon be joining the Asian Tigers, such as the "Mekong Six" countries (Cambodia, Vietnam, Laos, Myanmar, Thailand and Yunnan Province in China). In addition to certain regions in India, a similar development is expected in other areas around the world, e.g. South Africa, and parts of South and Central America.

Symptoms: enhanced greenhouse effect, local climate change, smog, acid rain, water pollution, health hazards, high consumption of resources.

Environmental Degradation Through Uncontrolled Urban Growth: Favela Syndrome

The *Favela Syndrome* refers to a process of unplanned, informal and thus environmentally harmful urbanization. Its features include various manifestations of poverty, such as the formation of slums and mostly illegal shanty towns. These are accompanied by overloading, infrastructural and environmental problems, as well as segregation of the population in terms of income, ownership, living standards and supply. As a result of increasing traffic and industrial emissions, and in the absence of adequate monitoring and regulatory requirements, air and noise pollution reach the highest levels in the world. Other aspects include the spread of surface sealing, uncontrolled accumulation of waste, sewage problems, and thus acute threats to the health of the population. Water is removed from peri-urban aquifers and conveyed to central districts (a symptom of the *Urban Sprawl Syndrome*), leaving poor people living in peripheral areas without an adequate water supply.

Factories in these cities fail to meet even minimum standards as far as safety precautions and measures to reduce waste and emissions are concerned. Vehicles are usually without any form of emission-reduction equipment. A typical feature of the *Favela Syndrome* is the predominance of the informal sector, since municipal planning and resource management is no longer able to maintain a basic infrastructure. The social and economic consequences are borne primarily by the urban poor.

Uncontrolled growth of settlements is a result of rapid population growth, on the one hand, and unresolved development problems in rural areas, on the other. Cities exert a "pull" effect on agricultural regions because they provide better opportunities for earning money. Furthermore, crop losses and declining productivity operate as highly effective "push" factors of rural-urban migration when the rural population is growing fast.

An example of this is Karachi, a megacity in Pakistan with approximately 7 million inhabitants that a century ago was still a village. Today, this city on the Indus delta is Pakistan's economic center, with two thirds of the country's industry concentrated here. The metropolis of Cairo, where 45% of the 12 million inhabitants live in informal settlements, also suffers from the *Favela Syndrome*. The number of people who live in cemeteries in Cairo is estimated at between 125,000 and 2 million. Other examples of cities with a pronounced disposition to the *Favela Syndrome* include São Paulo, Calcutta, Manila and Teheran. The areas exhibiting such a disposition are not confined to marginal peripheral zones of these megacities (e.g. mountain slopes that are difficult to develop, flood areas of rivers). In essence, the *Favela Syndrome* also includes settlement types that can be described as "no longer a village – not yet a city" (e.g. Nouakchott in Mauritania).

Symptoms: air pollution, soil erosion, accumulation of waste, noise, population growth, rural exodus, acute health hazards, socioeconomic marginalization, failure of public administration, lack of basic infrastructure, overloaded traffic infrastructure.

DESTRUCTION OF LANDSCAPES THROUGH PLANNED EXPANSION OF URBAN INFRASTRUCTURES: URBAN SPRAWL SYNDROME

The *Urban Sprawl Syndrome* refers to urban expansion with far-reaching environmental impacts. The formation of urban agglomerations (concentration and fusion of urban systems) leads to entirely new spatial structures and a need for corresponding adaptation. Agglomerations are characterized by high population density and specific, environmentally degrading network features which distinguish them clearly from other settlement structures.

In present-day Germany, natural habitats are being converted into useful infrastructural areas at a rate of 80-90 hectares a day. Thus one result, besides the well-known phenomena of soil compaction, surface sealing and fragmentation of habitats, is loss of biodiversity. It is estimated that about 30% of the plant and animal species in Great Britain have been permanently lost due to conversion of agricultural land.

A rise in traffic volume produces higher levels of direct soil pollution by automobiles through depositions from exhaust fumes, tire wear, oil residues, etc. Soils are further impaired by road traffic due to the damage caused to wayside vegetation by increasing immissions. When transport infrastructures are expanded to form major communication axes linking key places, this is generally accompanied by a capital-intensive and planned restructuring of the social infrastructure. This structural transformation not only degrades the environment, it also generates increased energy turnover (expansion of transport infrastructure correlates positively to the mobility of the population) and new material flows.

The syndrome is partly a consequence of reduced transportation costs and of infrastructure policies favoring the spread of built-up areas (the housing area per person, for example, rose in Germany from 15 m^2 to 34 m^2 during the period from 1950 to 1981). As a rule, it is only possible to meet this demand with new construction at the urban periphery. Parallel to this process, urban centers have displacement effects; price differentials between city centers and surrounding districts increase pressure on peripheral areas. The increasing separation of living and working manifests itself in particular through rising traffic volume.

The occurrence of the *Urban Sprawl Syndrome* does not preclude other structural malaises, such as the *Favela Syndrome*, operating in parts of settlements. In contrast to the *Favela Syndrome*, the *Urban Sprawl Syndrome* requires the existence of urban structures because growth impulses mostly emanate from the cities themselves; even when its population remains stable, the area of land covered by a city continues to grow.

The syndrome can be identified, for example, in the polycentric agglomeration of Los Angeles, which encompasses more than 120 incorporated cities of extremely varying magnitude (such as Santa Monica, Pasadena, or Long Beach). The agglomeration process in these cities was not only accompanied by large-scale infrastructural growth and the concomitant destruction of the environment; it also led to growing segmentation and polarization of the labor market and hence of socioeconomic structures. Changes of this kind have recently become apparent in the Ruhr area of Germany as well.

Symptoms: fragmentation of ecosystems, near-surface ozone contamination, stratospheric ozone depletion, urban air pollution, enhanced greenhouse effect, acid rain, soil contamination, compaction and surface sealing, health hazards, traffic congestion.

SINGULAR ANTHROPOGENIC ENVIRONMENTAL DISASTERS WITH LONG-TERM IMPACTS: MAJOR ACCIDENT SYNDROME

The central feature of the *Major Accident Syndrome* is the mounting threat to the environment in the form of singular localized disasters caused by humans, whereby liability for possible damage is limited or inadequate. Such improbable events with grave and often transboundary impacts, and the global perspectives for disaster prevention seem to be gaining in significance within the global change context. The rise in global transport-sector activities and increasing local demand for energy and raw materials enhance the risk of tanker accidents or, more generally, the incidence of environmental disasters in connection with the transport of hazardous goods. Other major threats besides the latter are industrial accidents. Equipment in newly-industrializing and developing countries is particularly vulnerable since the safety requirements there and their enforcement are less restrictive and appropriate disaster management is often lacking. Risks frequently stem from insufficient maintenance of industrial facilities. They include the large number of old nuclear power plants, chemical and other industrial plants in newly industrializing countries, transitional economies and developing nations that do not deploy best available technology.

The *Major Accident Syndrome* also embraces the introduction of non-indigenous species via the ballast water and hulls of ocean-going ships, with all the unforeseeable and sometimes catastrophic consequences this can have for other ecosystems. The resulting degradation may range from the extinction of indigenous species, the destruction of habitats through mass reproduction of invasive species, to endangering of the structure and function of entire ecosystems through the irreversible release of genetically modified organisms.

Examples of the *Major Accident Syndrome* are widely known due to the intense media interest in particular incidents and their aftermath. Names like Seveso, Chernobyl, Exxon Valdez and Bhopal are synonymous for this syndrome, and involved damage to humans and nature on a continental scale. The best-known example of introducing species into other habitats is undoubtedly the importing of rabbits to Australia; following extreme population growth and devastation of habitats, the measures used to combat the plague endangered indigenous fauna as well.

Symptoms: loss of biodiversity, ecosystem degradation, contamination of soil, water and air, health hazards.

2.2.3
"Sink" Syndromes

ENVIRONMENTAL DEGRADATION THROUGH
LARGE-SCALE DIFFUSION OF LONG-LIVED
SUBSTANCES: SMOKESTACK SYNDROME

This syndrome describes the remote effects of substance emissions following disposal in the environmental media water and air. The background to this problem is the failure of the strategy to dispose of undesired substances simply and easily by distributing them as finely as possible in the environment, or by diluting them greatly in environmental media (water, air). But air pollutants are not eliminated by high smokestacks – rather the problem is merely transplanted to other regions remote from industrial activity.

Depending on the emission patterns and the physical-chemical characteristics of the substances in the environmental media, the resultant distribution is local (typical for dust), regional (typical for NH_3, SO_2 and NO_x) or global (typical for CO_2, CFCs). Long-distance transport is primarily via the atmosphere or through flowing water bodies. Environmental impacts are distinguished according to whether the pollutants have effects on the system after dispersal in the environment (e.g. ozone depletion caused by CFCs, enhanced greenhouse effect due to CO_2 emissions), or whether they reaccumulate (acid enrichment in soils resulting from emissions of NH_3, SO_2 and NO_x, accumulation of persistent pesticides in the food chain).

The worldwide impacts of anthropogenic radiative forcing caused by emissions of greenhouse gases

(CO_2, CH_4, etc.) are an example of how the syndrome works. The alteration of the chemical composition of the Earth's atmosphere (especially by the use of fossil fuels) may be minimal in absolute terms, but has major implications for the Earth's radiative balance and thus for global climate.

A similar case concerns the depletion of the stratospheric ozone layer. Very small amounts of highly reactive substances (CFCs) cause an unexpectedly powerful and surprisingly large disruption of the chemical processes of the atmosphere, leading in turn to more UV-B radiation penetrating the atmosphere and the threats to human health and ecosystems that this involves.

Soil acidification provides another illustration of how the syndrome operates: emissions of the acidification agents SO_2 and NO_x, emanating above all from the energy and transport sectors, produce increased deposition of sulfuric and nitric acid in ecosystems, with subsequent acidification of soils. These processes are one of the main causes of forest decline in central Europe, a hitherto unknown phenomenon.

Symptoms: loss of biodiversity, eutrophication of ecosystems, depletion of the stratospheric ozone layer, increased levels of UV-B radiation falling on the Earth's surface, enhancement of the greenhouse effect, regional and global climate change, sea-level rise, acid rain, contamination of soils and groundwater with impacts on drinking water resources.

ENVIRONMENTAL DEGRADATION THROUGH
CONTROLLED AND UNCONTROLLED DISPOSAL
OF WASTE: WASTE DUMPING SYNDROME

The *Waste Dumping Syndrome* describes the growing need worldwide for controlled disposal of residual and waste matter. In contrast to the *Smokestack Syndrome*, where the underlying intention is to minimize pollution by "diluting" it in the air or water, this syndrome involves "localization", compaction and accumulation. Waste materials are collected in concentrated form at small-scale facilities and isolated from the environment as far as possible. Whereas over 50,000 waste dumps were still in use in Western Germany alone in 1970, only about 350-450 central large-scale landfills are to be operated in Germany in the future. Concentrating waste dumping in this way permits the deployment of sophisticated systems (e.g. landfill sealing, underground discharge systems, extraction facilities for landfill gas, intelligent monitoring systems).

In the final analysis, however, no one knows the precise lifetime of such systems as far as liquid or volatile pollutants are concerned; the stability of liners and the decomposition processes are acknowledged uncertainty factors. African agglomerations, for example, often have enormous landfill sites in their vicinity that can only be described as environmental "time bombs". In Southeast Asia, near the city of Manila, a gigantic waste tip has gained notoriety as "Smokey Mountain". Today, the syndrome can be found near large settlement areas on all the continents. Groundwater, drinking water, soils and air can be contaminated, depending on local environmental standards. Waste dumping also ties up financial and human resources for long periods of time, because it is only a matter of time before remediation and restoration work must be carried out. Dumping of hazardous radioactive waste is a special issue on account of the timescales involved – the storage facilities must remain totally isolated from the surrounding environment for several millenia, a technological and social challenge that has still not been resolved.

Symptoms: contamination of soils and groundwater, with harmful effects on drinking water, health hazards.

LOCAL CONTAMINATION OF ENVIRONMENTAL ASSETS AT INDUSTRIAL LOCATIONS: CONTAMINATED LAND SYNDROME

The *Contaminated Land Syndrome* characterizes sites and regions with accumulated depositions of pollutants in soils or underground that are a hazard to human health and the environment.

Contaminated sites can be found at locations and in regions where industrial, commercial or military activities used to occur, although they can also be found at abandoned and disused sites for storing solid municipal and industrial waste, or hazardous production residues.

Ecological, economic and social effects can overlap at sites were there has been an accumulation of different pollutants. This syndrome is primarily found in agglomerations where large-scale industrial plants were previously operated, e.g. in heavy industry, in the chemical industry and in the mining sector, and where, for a variety of reasons, hazardous waste disposal and other environmental aspects of production were given inadequate consideration.

One example of this syndrome is the agglomeration around (Saxony-Anhalt). Many other "hot spots" with this syndrome exist worldwide, for example Cubatao (Brazil), the Donez Basin (Ukraine), Katowice (Poland), Wallonia (Belgium), Manchester-Liverpool-Birmingham (Great Britain) and Pittsburgh (USA).

Symptoms: loss of biodiversity, deposition of pollutants in soils, water and air, loss of soil functions, health hazards.

2.3
Assignment of Core Problems of Global Changes to Syndromes

One advantage of basing the organizational structure for research on the syndromes of global change is that there is no longer an exclusive focus on problem areas within the various environmental media, but instead a common structure enbracing the problem area, its causes and its effects.

Each syndrome serves here as a research focal point, around which specialized disciplinary issues are then grouped. Furthermore, each of these patterns of global change generates, almost automatically, a number of cross-cutting issues that call *a priori* for interdisciplinary research strategies.

Such an approach can be implemented successfully only if all *core problems of global changes* can be identified within the syndromes. The assignment of core problems to the syndromes is carried out in *Tab. 3.*

The rows in this matrix correspond to the sixteen "clinical profiles" of the Earth System, while the columns represent those environmental problems of global significance as described in *Box 15.* The cells of the matrix are marked wherever a certain global problem contributes significantly to a given profile.

Tab. 3 shows that all core problems of global change bear a relation to several syndromes at once. Quite obviously, syndromes are cross-sectoral phenomena each encompassing key causal factors, the mechanisms by which they operate, and major contributions to core problems in a significant manner. The first row of the matrix, representing the *Sahel Syndrome*, illustrates the contrasting weaknesses inherent in the traditional approach, which analyzes problems from too narrow a perspective – here, critical trends such as "soil degradation" or "migration" manifest themselves as both causes *and* effects of the *Sahel Syndrome*, while loss of biodiversity is predominantly a resultant impact of the syndrome in question.

From the outset, syndrome analysis returns such factors to their rightful place in the causal (feedback) chain. This *systems-based* method of analysis and description provides an organizational framework for deciphering the bewildering complexity of global change in nature and human society. If research is organized according to the main syndromes of the Earth System, then existing resources can be focused on problems *and* causes, thus enhancing the efficiency of research activities.

Table 3
Assignment of core problems of global changes to syndromes.
Source: WBGU

Syndrome \ Core problems	Climate change	Loss of biodiversity	Soil degradation	Scarcity and pollution of freshwater	Threats to world health	Threats to food security	Population growth and distribution	Man-made disasters	Overexploitation and pollution of the world's oceans	Global disparities in development
Sahel Syndrome		•	•	•		•	•	•		•
Overexploitation Syndrome	•	•	•	•				•	•	•
Rural Exodus Syndrome		•	•			•	•	•		•
Dust Bowl Syndrome	•	•	•	•		•		•		
Katanga Syndrome		•	•	•						
Mass Tourism Syndrome		•	•	•				•		
Scorched Earth Syndrome		•	•		•	•	•			•
Aral Sea Syndrome	•	•	•	•			•	•		•
Green Revolution Syndrome		•	•	•	•	•	•			•
Asian Tigers Syndrome	•	•	•	•	•		•			•
Favela Syndrome	•		•	•	•		•			•
Urban Sprawl Syndrome	•	•	•	•						
Major Accident Syndrome		•	•		•					
Smokestack Syndrome	•	•	•		•	•		•		
Waste Dumping Syndrome		•	•		•					
Contaminated Land Syndrome		•	•		•				•	

Designing and planning environmental research require that consideration be given to such generally applicable criteria as
- quality of expertise (especially methodological quality),
- competitiveness within a national and international comparison,
- cost-benefit aspects,
- "visibility",
- relevance to specific application, etc.

However, global change is not a research topic like any other: the existential importance of these problems for the future development of humankind, and the uniqueness, complexity, variety and dynamics of the phenomena involved, make it necessary to deploy a number of other relevance criteria for research policy. Putting these criteria into operation can fulfill a dual purpose – orienting research activity to the cross-sectional character of environmental issues, and achieving more efficient prioritization when financial resources are scarce.

The Council recommends that the following criteria in particular be applied in Germany when selecting research topics in the field of global (and regional) change:

R1: GLOBAL RELEVANCE
Are key parameters, basic patterns or core problems in the Earth System being investigated? Are large numbers of people affected by the problem? Is the research likely to generate new options for controlling the environment and development process?

R2: URGENCY
Are answers needed quickly in order to prevent irreversible environmental or socioeconomic developments with severe negative outcomes?

R3: GAPS IN KNOWLEDGE
Can serious gaps affecting a holistic view of the global environment and its dynamics be closed?

R4: RESPONSIBILITY
Are problems being investigated for which Germany is directly or indirectly responsible (e.g. through greenhouse gas emissions or as a participant on the world market for goods and services)? Does the topic relate to general ethical principles (e.g. preservation of life on earth)?

R5: NATIONAL IMPACT
Are problems being researched which could have direct or indirect effects on Germany (e.g. impacts on climate, environmental refugees)?

R6: RESEARCH AND PROBLEM-SOLVING COMPETENCE
Does the research relate to areas where Germany contributes substantially on account of its scientific, technological and infrastructural potential? Can research on the topic lead to further improvement of that potential and thus to enhancement of Germany's attractiveness for investment?

4 Integration Principles

As a basic principle, the analysis of global change should focus on existing problems and potential applications. Achieving a global perspective requires collaboration between and the integration of different disciplines, interest groups and actors. The diversity of concepts for communicating environmental knowledge means that many problems must be overcome for such integration to occur. The key issue for researchers concerns the principles according to which the requisite synthesis is to be realized.

In the section below, the Council has compiled a number of principles that may prove helpful in the implementation of integrated environmental research. The integration of research activities should, if carried out according to these principles, be based on analytical, methodological and organizational aspects, and be geared to certain implementation aspects.

4.1
Analytical Integration Principles

I1: SPATIAL REFERENCE
This criterion calls for collaboration in relation to a common geographical area. Such collaboration may be achieved with the syndrome concept, for example, which describes and explains "clinical profiles" in terms of geographical overlapping of specific trends. Syndrome-oriented research should bring the individual disciplines together to address a specific geographical area.

I2: TEMPORAL REFERENCE
Another dimension is that of time. If one considers that many global environmental problems result from ecosystems being unable to adapt quickly enough to environmental changes, and that economic and social adaptation processes require varying lengths of time, then cooperation between the various disciplines should feature a common temporal perspective.

I3: SOCIOCULTURAL STRUCTURES AND PROCESSES
Global society is structured into sub-societies each with its own level of development, education level and value system – sociocultural structure, in other words. This affects factors such as willingness to accept risks, adaptive capacity, sensitivity to environmental issues, and environmentally relevant behavior. When integrating research activities, these differences should be explicitly addressed in order to improve the efficiency of research and its subsequent implementation.

4.2
Methodological Aspects

I4: MODELING AND SIMULATION
Every model is an attempt to simulate reality by means of simplified hypotheses that take account of interdependencies and which should be validated with empirical data. This means that constructing models is a suitable way to integrate disciplines, whereby simulation focuses attention on those hypotheses that are either in need of critical review or which reveal gaps in the overall model.

I5: JOINT INSTRUMENTS
Joint instruments further integration in the same way. These may involve the coordinated (complementary) use of large-scale equipment (e.g. satellites, research ships or supercomputers), infrastructures and knowledge resources (e.g. databases or algorithms).

4.3
Organizational Aspects

I6: INTERDISCIPLINARY FACILITIES
This method of integration centers on creating research institutions with clearly defined cross-cutting foci, which therefore necessitates collaboration between natural scientists, engineers, economists and

social scientists as appropriate, depending on the specific task at hand.

I7: TEMPORARY ASSOCIATIONS

These refer to the formation of medium-term, project-based networks between established institutions of specific disciplines in order to promote integration. Where necessary, authority should be assigned to joint management committees whose members are linked through modern communications media ("data highways").

I8: SUPPORT STRUCTURES AND PROGRAMS

The main elements here are the establishment by the federal government of interdepartmental priority programs of a cross-sectional nature (e.g. Migration Research, Health and Global Change), the strengthening of interdisciplinary research networks by the DFG (e.g. reorganization of research assessment, promotion of collaborative research centers treating a specific complex of topics, rather than methodically defined and geographically distributed centers), the establishment of Max Planck Institutes focusing on environmental processes (e.g. for studying global biogeochemical cycles) or instituting prizes for environmental research that achieves a "synthesis" between different disciplines.

I9: ORIENTATION TO INTERNATIONAL PROGRAMS

Greater involvement in international programs (e.g. Framework Programmes of the EU, international programs on global change, joint *capacity building* activities in developing countries) is one way to overcome the lack of global perspectives in environmental research. Another benefit of such collaboration is that it promotes the growth of international research networks.

I10: EDUCATION AND TRAINING

Integration is furthered by setting up foundation-level and advanced-level courses in environmental subjects (e.g. agricultural ecology, geoecology, environment and business, systems analysis of human/environment relations), graduate colleges, summer schools and academic exchange programs.

4.4
Implementation Aspects

I11: PARTICIPATION

Integrative effects can be achieved through greater participation on the part of those causing, affected by and combating environmental problems in the environmental research process. Potential partners are municipalities (e.g. climate protection alliances), lobbies, industry (e.g. the energy industry, or the insurance and reinsurance sector) and groups active in environmental politics.

I12: EVALUATION

Experience shows that evaluation of research can promote integration. Evaluation should address the question as to what kind of research is needed in order to understand syndromes and to manage them through elimination of their non-sustainable components.

5   Syndrome Ranking

The Council has conducted an internal expert survey based on the relevance criteria outlined in *Section C 3*. The aim was to identify those syndromes which the German research community should prioritize. Each of the 16 syndromes was evaluated according to the relevance criteria. The experts were also asked to assess their own competence concerning the various complexes in order to ensure that the relevance of each syndrome was evaluated as objectively as possible. This was necessary because the Council members are specialized in different areas. The survey participants were also given an opportunity to reserve judgment on particular syndromes. The opinions on syndrome relevance were weighted during compilation according to the expert's assessment of his or her competence. The results of the survey are shown in *Tab. 4*, which groups assessments according to the six relevance criteria.

On the basis of these results, the syndromes were then classified into three categories of priority, whereby all the relevance criteria were weighted equally (*Tab. 5*). The syndromes in each category were not ranked any further, and are listed alphabetically. The evaluation identified seven problem complexes as having uppermost priority (*Section D 3*).

The initial rough assessment of the syndromes to be prioritized for research should be debated in a special discourse on the syndrome concept among global change researchers and decision-makers, and modified where necessary. The Council proposes that a methodically prepared Delphi study be carried out among a larger group of experts (*Section D 3*).

R_1 Global relevance		R_2 Urgency		R_3 Gaps in knowledge	
Smokestack S.	3.9	Sahel S.	3.9	Aral Sea S.	2.8
Sahel S.	3.7	Smokestack S.	3.8	Favela S.	2.8
Favela S.	3.7	Favela S.	3.8	Urban Sprawl S.	2.8
Overexploitation S.	3.6	Overexploitation S.	3.6	Scorched Earth S.	2.5
Waste Dumping S.	3.3	Rural Exodus S.	3.4	Rural Exodus S.	2.5
Urban Sprawl S.	3.1	Waste Dumping S.	3.3	Mass Tourism S.	2.5
Green Revolution S.	3.0	Urban Sprawl S.	3.1	Contaminated Land S.	2.5
Rural Exodus S.	2.8	Katanga S.	3.1	Waste Dumping S.	2.5
Contaminated Land S.	2.8	Mass Tourism S.	3.1	Overexploitation S.	2.4
Katanga S.	2.8	Green Revolution S.	3.0	Sahel S.	2.4
Mass Tourism S.	2.8	Dust Bowl S.	2.9	Oilspill S.	2.3
Scorched Earth S.	2.7	Scorched Earth S.	2.8	Smokestack S.	2.2
Dust Bowl S.	2.7	Asian Tigers S.	2.8	Asian Tigers S.	2.1
Asian Tigers S.	2.6	Contaminated Land S.	2.8	Green Revolution S.	2.1
Aral Sea S.	2.5	Aral Sea S.	2.7	Katanga S.	2.0
Oilspill S.	2.2	Oilspill S.	2.5	Dust Bowl S.	1.8

Table 4
Ranking of syndromes on the basis of the relevance criteria (*Section C 3*). The table shows the results of an internal survey of Council members. Relevance was assessed on a scale of 1 (low) to 4 (high).
Source: WBGU

▶

Table 4
Continued

R₄ Germany's responsibility		R₅ National Impact on Germany		R₆ Germany's research and problem-solving competence	
Smokestack S.	3.5	Smokestack S.	3.5	Smokestack S.	3.9
Mass Tourism S.	3.5	Dust Bowl S.	3.1	Waste Dumping S.	3.6
Dust Bowl S.	3.3	Contaminated Land S.	3.1	Contaminated Land S.	3.4
Contaminated Land S.	3.2	Waste Dumping S.	3.0	Katanga S.	3.2
Waste Dumping S.	3.2	Urban Sprawl S.	2.9	Green Revolution S.	3.2
Katanga S.	2.8	Katanga S.	2.9	Dust Bowl S.	3.2
Urban Sprawl S.	2.7	Mass Tourism S.	2.7	Aral Sea S.	3.2
Overexploitation S.	2.5	Scorched Earth S.	2.5	Oilspill S.	3.2
Oilspill S.	2.4	Oilspill S.	2.3	Urban Sprawl S.	3.1
Green Revolution S.	2.4	Sahel S.	2.3	Sahel S.	3.0
Aral Sea S.	2.2	Favela S.	2.3	Overexploitation S.	2.9
Sahel S.	2.1	Rural Exodus S.	2.1	Scorched Earth S.	2.9
Rural Exodus S.	1.8	Aral Sea S.	2.0	Asian Tigers S.	2.9
Scorched Earth S.	1.8	Green Revolution S.	1.6	Mass Tourism S.	2.6
Favela S.	1.7	Overexploitation S.	1.6	Favela S.	2.6
Asian Tigers S.	1.6	Asian Tigers S.	1.6	Rural Exodus S.	2.5

Table 5
Classification of
syndromes according to
priority categories.
German research should
prioritize the seven
syndromes in Category 1.
Source: WBGU

Category I	Category II	Category III
Contaminated Land S.	Aral Sea S.	Asian Tigers S.
Dust Bowl S.	Favela S.	Oilspill S.
Mass Tourism S.	Green Revolution S.	Rural Exodus S.
Sahel S.	Katanga S.	Scorched Earth S.
Smokestack S.	Overexploitation S.	
Urban Sprawl S.		
Waste Dumping S.		

6 Designing a Syndrome-based Research Structure: A Case Study on the Sahel Syndrome

The aim in this section is to demonstrate how a research structure can be designed which is able to identify the interdependencies between syndrome-specific problems and to elaborate appropriate response strategies. The individual steps are first discussed in general terms before an illustrative example is provided on the basis of the *Sahel Syndrome* (for a description *see Section C 2.2.1*). The proposed procedure can be applied analogously to the other syndromes of global change.

Basing research on the syndrome approach means, first and foremost, that its methods (*Section C 2*) be used for a research strategy that identifies the pertinent issues and ensures that the most efficient methods of addressing them are applied. Such a strategy is composed of the following elements:

- The *network of interrelations* for the specific syndrome, displaying its critical trends, driving forces, impacts and mechanisms.
- The *disposition space* by means of which regions affected by a syndrome and – even more important for research strategy – the vulnerability of a region can be identified. The ecological and socioeconomic *disposition factors* play a major role with respect to those research fields of relevance for the future. Together, these two steps form the core of a qualitative systems analysis of the syndrome, the results of which can be used to formulate syndrome-specific questions for global change research.
- Global change research based on the syndrome concept must also be organized in line with certain *relevance criteria* (*Section C 3*) and *integration principles* (*Section C 4*) so that the financial and scientific resources are used in the most targeted manner possible. When these aspects are taken into account, systems analysis can be translated into a specific way of organizing research activities, for which recommendations are developed in the course of this section.

The above procedure will be applied below to the *Sahel Syndrome*. Research issues will be derived from the network of interrelations (*Section C 6.1*) and the disposition space (*Section C 6.2*), following

which we shall propose ways of addressing these issues in accordance with the relevance criteria and integration principles (*Section C 6.3*). Finally, we put forward a model for research organization containing the basic structure of a network for researching the *Sahel Syndrome* (*Section C 6.4*).

Our recommendation that a research network be established results from an overall assessment which says that, although global change research is a major task for the future and involves new challenges, it is by no means necessary to start "from scratch" everywhere. The research landscape at international level and within Germany is sufficiently diversified to perform these tasks, provided that targets, organization and frameworks are adequate. What is often lacking is cooperation, communication and integration, as opposed to good approaches and actual research results. This is where the network idea comes in. To illustrate this, we detail a number of institutions and projects in which research work is already being carried out on issues relating to the *Sahel Syndrome*, and suggest ways of enhancing their integration. These proposals are merely examples to show how the syndrome concept can be translated into research organization. Our intention is not to examine and evaluate German research on the Sahel in any exhaustive fashion, but rather to demonstrate that the syndrome concept is a suitable means for evaluating existing research and integrating it into a new strategy.

6.1
The Network of Interrelations for the Sahel Syndrome

To describe the pattern of interactions within a syndrome, the Council has developed an instrument (see *Section C 2.1* as well as WBGU, 1993 and 1995a) in which trends and interrelations are presented in graphic form. With the help of this syndrome-specific *network of interrelations*, it is possible to describe the relevant interactions and obtain a comprehensive and clear picture of the syndrome. The *Sahel Syndrome* refers to the complex web of causes and effects associated with the agricultural overexploitation of marginal land (WBGU, 1995a) and is described in greater detail in *Section C 3.2.1*. The central mechanism of the *Sahel Syndrome* is the mutual reinforcement of environmental degradation, social and economic marginalization and overexploitation. *Fig. 7* depicts this *vicious circle* and shows the relevant trends and interactions along the people-environment interface. The critical aspect, however, is that this vicious circle is not some isolated structure, but is interrelated to numerous other trends of global change. The syndrome-specific network of interrelations, i.e. the overall pattern of trends and their interrelations, is shown in *Fig. 8*. Connecting lines with arrows at the end symbolize a reinforcing while those with solid circles indicate an attenuating interaction. The black connecting lines designate interrelations that are of significance for the syndrome, but play a subordinate role in deriving the three core research issues (see below). Trends regarded as important for the *Sahel Syndrome* but as secondary in the global context are depicted in a rhombus-shaped box (*Fig. 8*).

The *central vicious circle* highlights the precarious social and ecological situation faced by large population groups, particularly in developing countries. It is essentially a dilemma: on the one hand, there is the need to ensure food security for the local population, which due to the lack of economic alternatives can only be achieved through intensification or expansion of agriculture or overexploitation of vegetation, both in the short and medium term. The question of response options on the part of the affected population under existing conditions is of special importance in this context. The other horn of the dilemma is the risk of greater soil degradation caused by inappropriate land management methods at marginal locations. Human-induced degradation of the environment has feedback effects on society, usually hitting vulnerable groups the worst and aggravating their situation, thus reinforcing the core mechanism of the *Sahel Syndrome*.

In addition to the vicious circle described above, one can identify three other structural elements, or sub-networks, that play a key role within the *Sahel Syndrome* :
- *The response options of the affected population* are severely restricted in the *Sahel Syndrome*. These constraints are intimately linked to the vicious circle. In many cases, the only way out for those concerned is to migrate to other regions or to the urban agglomerations. Population pressure and advancing poverty mutually reinforce each other (sub-network of interrelations shown in red in *Fig. 8*).

- *Regional climate change* is a complex of interactions in which the conversion of natural ecosystems brings about a change in local, and in some cases global climate. This climate change can in turn have significant impacts on the water resources of the region in question. What makes this interaction important is that it unfolds over long periods of time and thus imposes a certain inertia on the syndrome's dynamics (sub-network depicted in green in *Fig. 8*).
- *Economic conditions* (both national and international) play a critical role in the *Sahel Syndrome*, triggering or accelerating its central mechanism (sub-network shown in blue in *Fig. 8*).

Although the *Sahel Syndrome* is a widespread and indeed typical feature of many developing countries, it also harbors germs of improvement and latent potential to disrupt the central vicious circle. Prime examples are the attenuating effect on population growth exerted when women enjoy a better position in society, or the transfer of adapted technologies (e.g. land management methods which preserve soils, use of energy-saving technologies), depicted in *Fig. 7* as attenuating trends. Reference will be made to these factors as well in the analysis that follows.

6.2
Disposition to the Sahel Syndrome

The network of interrelations is a graphic portrayal of the basic pattern of people-environment relations within a syndrome. Disposition towards a particular syndrome, on the other hand, defines the probability of an "outbreak" of its specific mechanisms. Ascertaining this disposition is thus the second step in the systems analysis of a syndrome.

A region is susceptible to a syndrome if certain structural constellations of disposition factors are present in the ecosphere and anthroposphere. Such factors include cultural characteristics, water availability, angle of slope, etc., and indicate a region's susceptibility when a syndrome-specific combination of these factors is present. If certain triggers (exposition factors such as substantial exchange rate fluctuations or periods of drought) begin to operate in this context, the mechanism of the syndrome is unleashed, and the syndrome becomes acute in the respective region.

The disposition to the *Sahel Syndrome* is determined primarily by the
- *natural dimension of disposition*, i.e. production conditions within the natural environment indicate that soil degradation is already liable to occur, albeit to a lesser extent, as a result of agricultural intensification or expansion;

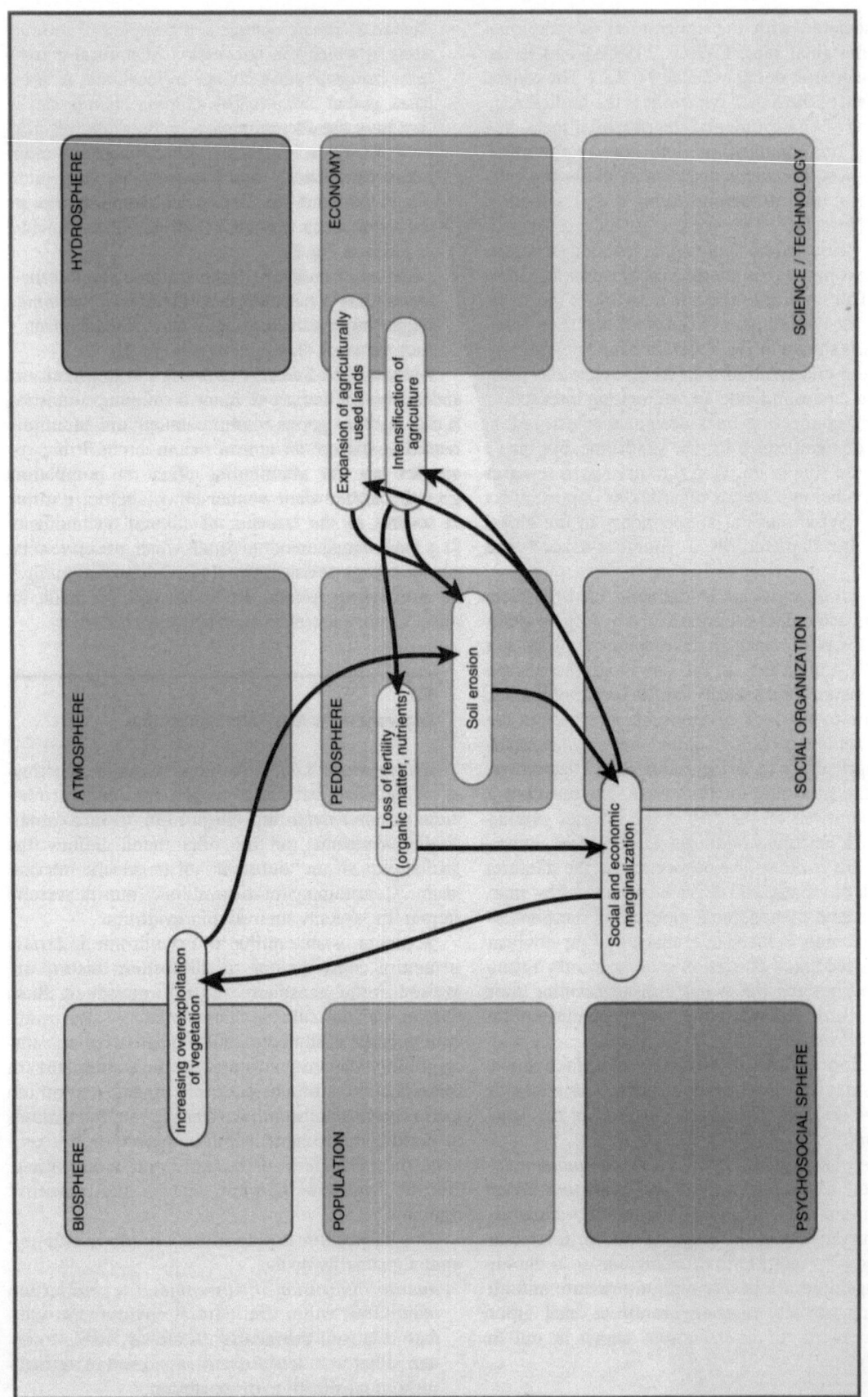

Figure 7
Central mechanism of the *Sahel Syndrome* (vicious circle).
Source: WBGU

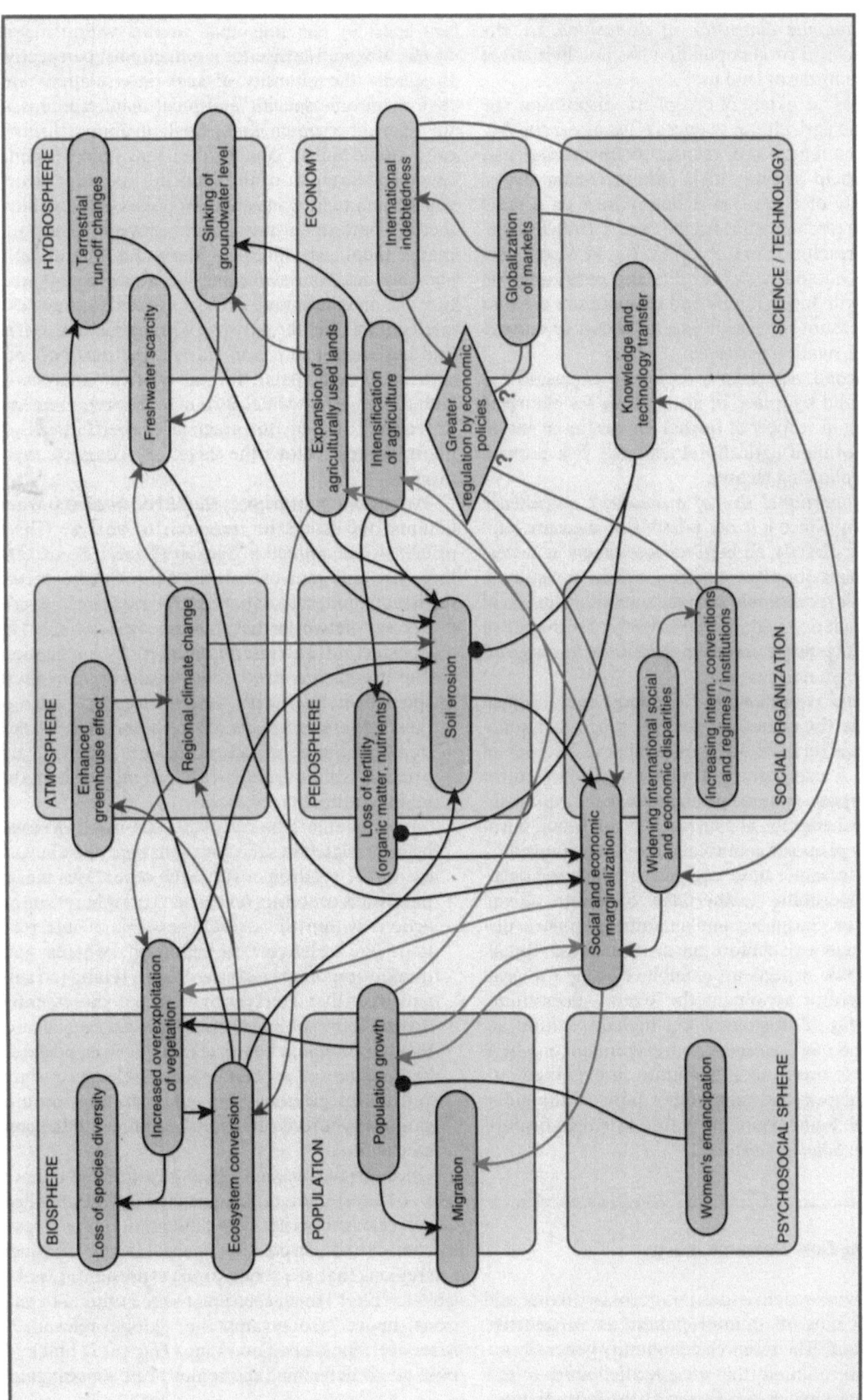

Figure 8
Syndrome-specific network of interrelations of the *Sahel Syndrome*. The three sub-networks from which the complexes of issues are derived are marked red, green and blue.
Source: WBGU

– *socioeconomic dimension of disposition*, i.e. the marginalized rural population has no alternatives to these forms of land use.

To assess the extent of ecospheric disposition, the natural and agricultural sciences must assess the fragility of the region with respect to agricultural use. With the help of fuzzy logic (Zimmermann, 1994), the fragility of a location is determined on a scale between 0 (certainly not fragile) and 1 (fragile location) at a resolution of 0.5° x 0.5° (*Fig. 9*). Slope, precipitation variability, soil fertility and constraints on plant growth due to aridity and temperature are taken into account here, as is ease of access to surface water for irrigation purposes.

The second aspect, socioeconomic disposition, is characterized by a lack of alternatives for the rural population in respect of further expansion or intensification of their agricultural activities. It is derived from the following factors:

* The *proportional size of a country's subsistence economy*. Since it is not possible to measure subsistence directly, an indirect assessment is necessary. This is done by comparing official statistics on domestic food supply with the nutritional needs of the population to derive an index for the degree of self-sufficiency, a rough indicator for the size of the subsistence economy.
* Boundary conditions in the energy sector which promote the syndrome, the key aspect being dependence on fuelwood in particular as a source of energy. A composite indicator was developed for this purpose, comprising an index of per capita energy consumption and another for fuelwood burning as a proportion of total energy consumption.

Fig. 10 indicates those regions with high socioeconomic vulnerability to the *Sahel Syndrome*. Given that both an ecospheric and an anthropospheric disposition must exist before the syndrome can "break out", the two aspects are combined using a logical AND operator to obtain the overall disposition, shown in *Fig. 11*. Regions with a high disposition, i.e. where the *Sahel Syndrome* is already found or where it is liable to break out in the future, are marked red. The map shows that many other parts of the globe besides the Sahel region itself display a high disposition to the *Sahel Syndrome*.

6.3
Deriving Core Research Issues

To derive research issues, it is necessary to use and "read" the network of interrelations as a productive heuristic tool. The research community generally examines interrelations that are specified more or less distinctly and which are shown at the most elemen-

tary level by the individual arrows within a syndrome. It would suffice for a reductionist perspective to specify the intensity of such interrelations and their syndrome-specific and local manifestations as the relevant research issues. Such an approach, however, would fail to capitalize on the special significance and strength of the syndrome concept, which consists in precisely its ability to take complex transsectoral patterns of interaction between trends and render them accessible to systems and interdisciplinary analysis. For a syndrome-based approach, therefore, the main task is not to pick out and examine isolated links within the network of interrelations, but to find key *trend-interaction clusters* that not only are sufficiently aggregated for the systems analysis of global change, but which also meet the requirements of research strategy for practicable specifications of interrelations below the level of complete syndromes.

Two essential principles should be observed when defining the issues for research to analyze. These principles are applied in *Sections C 6.3.1 - C 6.3.3* to three core sub-networks of the *Sahel Syndrome* and the issues connected with them. These principles are:

* The sub-networks that structure research must be transsectoral and interdisciplinary by implication. The trends involved must span two or more spheres of the Earth System. Because each of these spheres corresponds in the main to a traditional discipline in natural science, applying the principle will invariably result in intense horizontal integration of research.
* Taken together, the individual sub-networks must be interrelated in such a way that they link key areas of the syndrome with each other. This means that issues centering on certain "peripheral" interactions within the overall network of interrelations are relatively unimportant, whereas extremely important issues are those relating to sub-networks that are synonymous for the essential driving forces of key trends. This criterion ensures that the complexes of issues thus derived possess a high degree of problem-solving relevance – and vertical integration – since the measures for mitigating a syndrome can only be applied to the core mechanisms.

Once conclusions have been drawn from the network of interrelations, the next step towards application of relevance criteria and integration principles is to consider the disposition space. Having identified the regions that are prone to the syndrome, it is possible to "filter" them according to the criteria of "national impact", for example, or "global relevance". Moreover, the disposition maps (*Figs. 9-11*) make it possible to determine the regions to be investigated (integration principle I_1), or regions with similar sociocultural structures (I_3) (*see Section C 4*). Thus, disposition can also be used to narrow down particular issues derived from the network of interrelations.

As an example, the *Sahel Syndrome* can be assessed as a whole in terms of "national impact" (R_5, *Section C 3*). It can be readily seen from *Fig. 11* that some of the regions with high disposition have close political or economic links with Germany or the EU – northern Africa and central Asia being cases in point. Furthermore, the maps can be used to identify those regions that require investigation (integration principle I_1) or which bear similarities with respect to the relevant sociocultural structures and processes (I_3) (*Fig. 10*).

Proceeding from the network of interrelations (*Fig. 8*) and the maps of the various disposition zones, the Council has identified three key complexes that will serve as examples for the derivation of syndrome-based research projects. The following three *complexes of issues* are elaborated from the core elements of the syndrome as discussed above, i.e. the central vicious circle and sub-networks marked in different colors in *Fig. 8*.

1. *Response options of the affected population* (marked red in *Figs. 8* and *12*): What response options are open to groups threatened by marginalization and what factors influence the adaptation or avoidance strategies they actually choose?
2. *Regional climate change*: (marked green in *Figs. 8* and *12*): What are the specific interactions between anthropogenic changes in the world·climate, regional climate, regional ecosystems and agricultural production systems, and local water availability in regions with a disposition to the *Sahel Syndrome*?
3. *International economic conditions* (marked blue in *Figs. 8* and *12*): What connection exists between trends and structures in the world economy, in national policy and in the socioeconomic marginalization processes in regions with a disposition to the *Sahel Syndrome*?

Tab. 6 provides an overview of the relevance criteria assigned by the Council to these three issues (*Section C 3*) and the integration principles necessary for handling the latter (*Section C 4*).

Elements of the sub-mechanisms highlighted here are also referred to in the *Desertification Convention*, which specifies research priorities for this complex of issues (for detailed references to the Desertification Convention *see* WBGU, 1995a). The Convention contains explicit reference to the socioeconomic causes of degradation, but the research projects specified in Article 17 are almost exclusively sectoral in nature, and the selection of topics does not pay due regard to

the complexity of interactions between ecospheric, cultural, social and political processes.

6.3.1
Complex 1: Options for the Affected Population

Marginalized groups generally have a restricted action space. Within the *Sahel Syndrome*, they are typically faced with a choice between migration, on the one hand, and overexploitation of vegetation and intensification or expansion of agriculture, on the other. Regions afflicted by the syndrome usually exhibit both of these response patterns. As can be seen from the disposition map, therefore, this field of research is not only of *global relevance* (R_1), but also of great *urgency* (R_2), both for those directly affected as well as for possible destination countries of migration flows, including, of course, the EU and Germany (R_5). Moreover, the vicious circle at the core of the Sahel syndrome is one of the central driving forces behind global population growth: faced by degraded resources, having many children can reduce the daily work load (e.g. gathering fuelwood, fetching water) and certainly helps to provide for old age in countries without social security systems. In this context, the research community has the task of designing response options, improving their communication to those concerned, or enabling them in the first place. Furthermore, research must ascertain how existing conditions affect the actual decisions taken by individuals. This is of immense importance, especially for prediction.

These issues involve various *gaps in knowledge* (R_3): Is there regional development potential by which intensification, overexploitation and migration pressures could be mitigated? What social groups are prone to marginalization, and what migration processes are triggered? What role does population growth play? What are the impacts of enhancing the role of women in society? How are options for the population likely to develop in different scenarios? To answer these questions, sectoral research projects involving the economic sciences and geography must be supplemented by work in the fields of sociology, psychology, social anthropology and related disciplines.

The overarching research issue – response options of the affected population – can be differentiated on the basis of three aspects that can be examined in parallel and which are linked to various research structures and programs already in existence:

- *Assessment of the options and strategies of marginalized population groups in regions afflicted by the Sahel Syndrome.* Here it is necessary to collate current research and existing results from region-

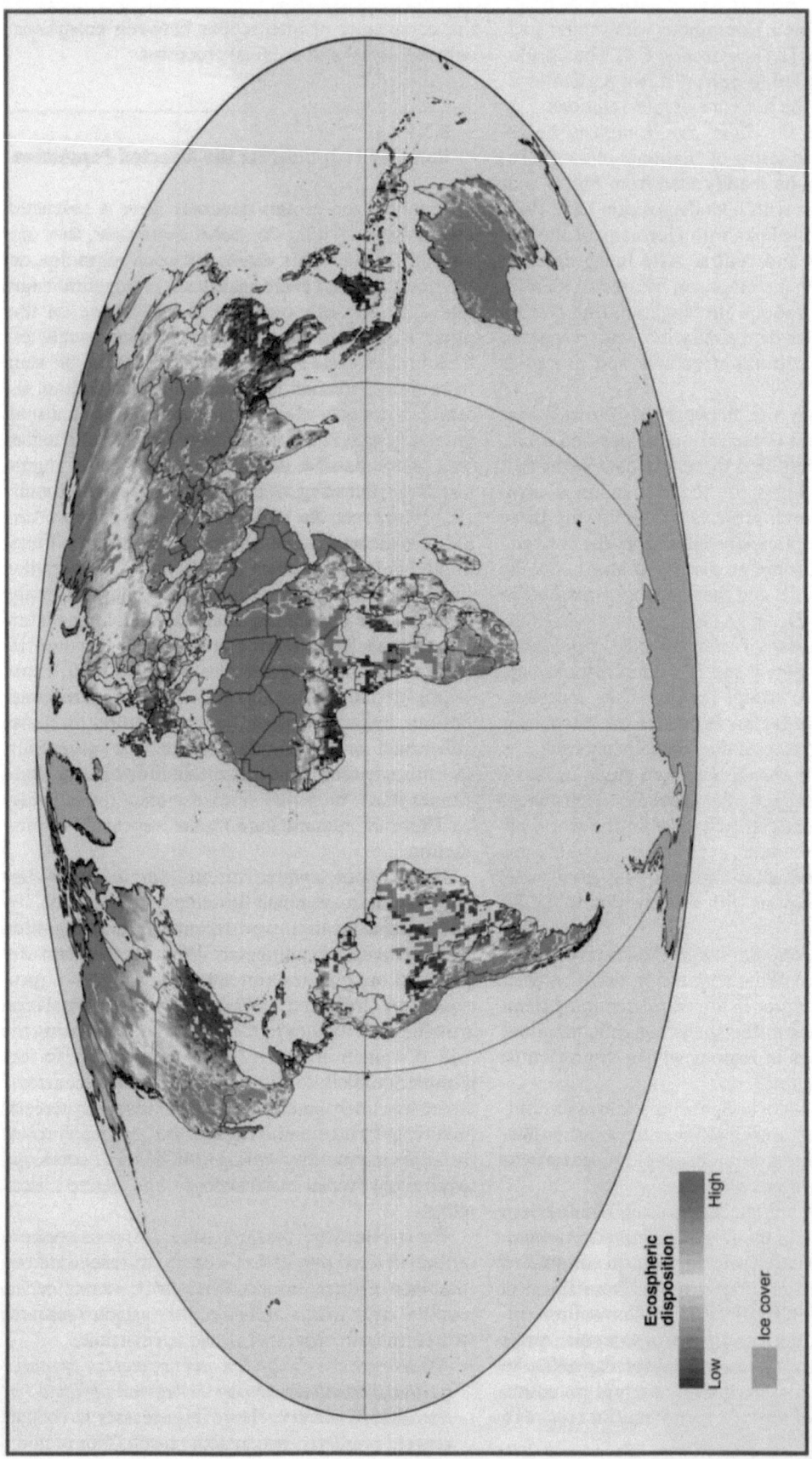

Figure 9
Ecospheric components of the *Sahel Syndrome* disposition space.
The fragility of a location's agricultural production due to natural factors is shown on a scale between 0 (green, certainly not fragile) and 1 (red, certainly fragile); factors include the angle of slope, variability of precipitation, constraints on plant growth due to aridity and temperature, soil fertility and ease of access to surface water for irrigation purposes. Large areas covered with glaciers are marked gray (equal-area Mollweide projection).
Source: BMBF-Project "Syndromdynamik" / PIK-Core Project QUESTIONS

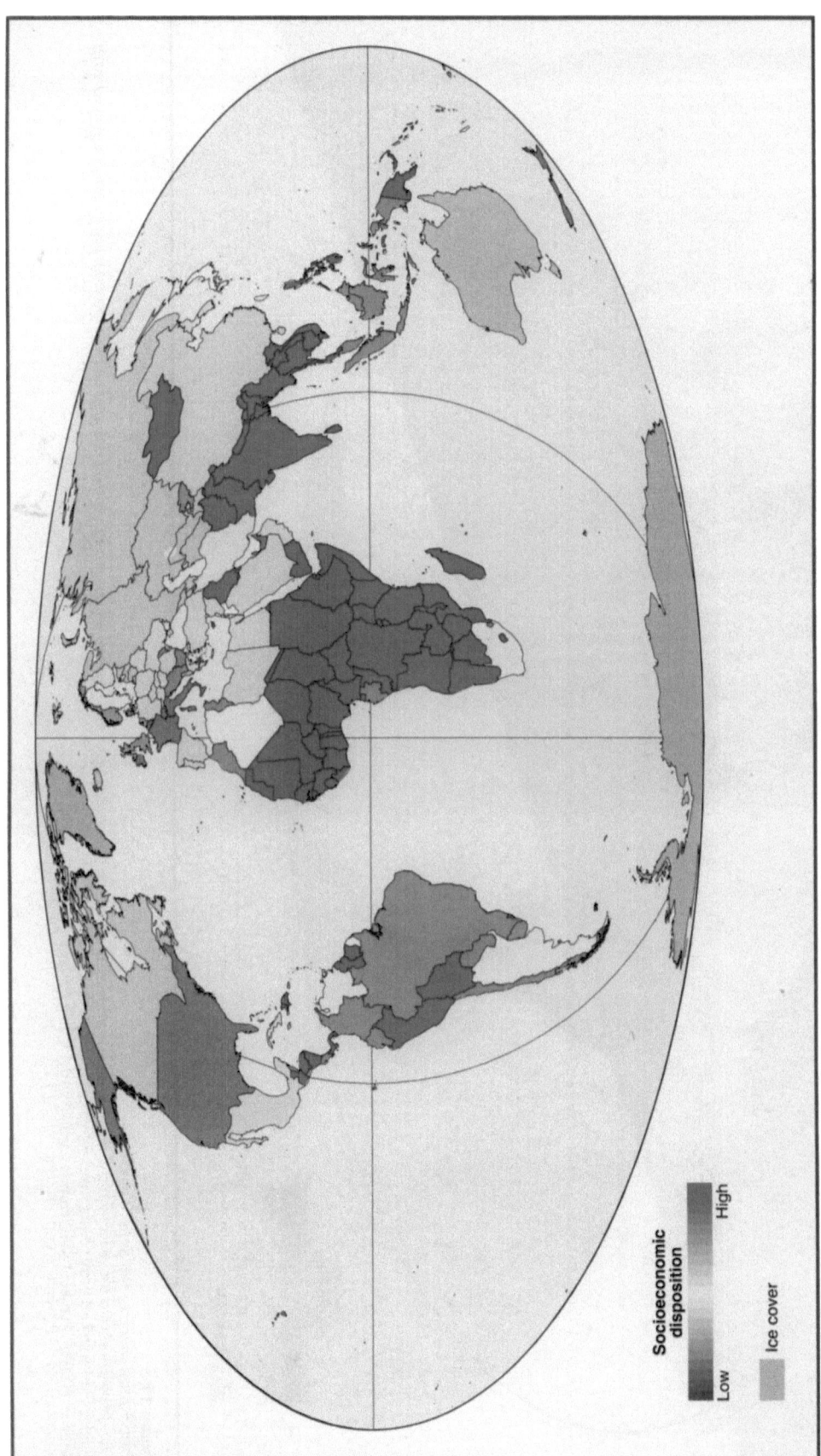

Figure 10
Socioeconomic components of the *Sahel Syndrome* disposition space.
The subsistence economy of a location is shown on a scale between 0 (green, certainly not a subsistence economy) and 1 (red, certainly a subsistence economy); the food situation based on market statistics in comparison to the food needs and a typical subsistence energy consumption pattern have been taken into account. Large-scale areas covered with glaciers are marked gray (equal-area Mollweide projection).
Source: BMBF-Project "Syndromdynamik" / PIK-Core Project QUESTIONS

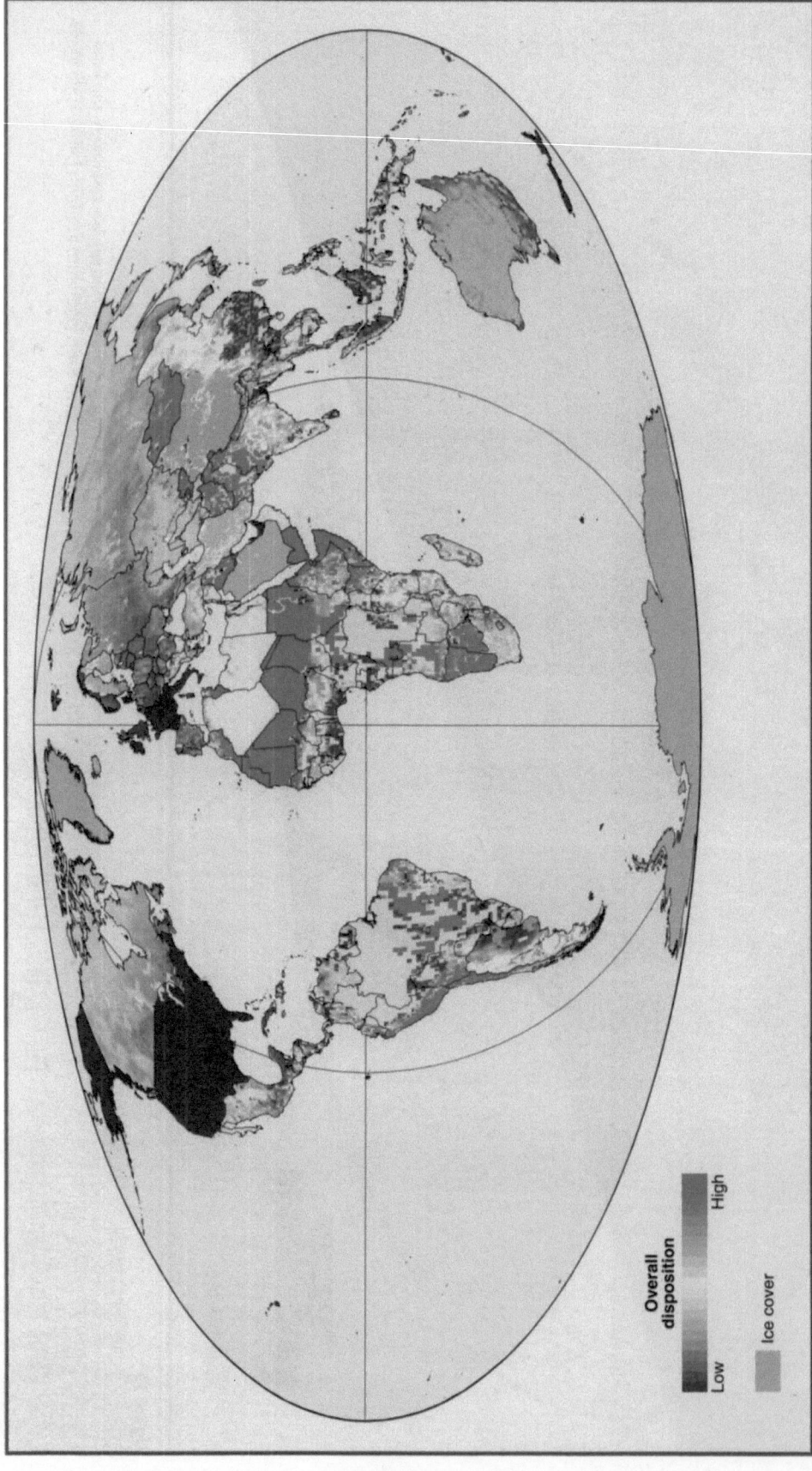

Figure 11
Disposition space of the Sahel Syndrome.
The disposition of a location to the *Sahel Syndrome* is shown on a scale between 0 (green, certainly not disposed) and 1 (red, certainly disposed). Large areas covered with glaciers are marked gray (equal-area Mollweide projection).
Source: BMBF-Project "Syndromdynamik" / PIK-Core Project QUESTIONS

Table 6

Selected complexes of issues on the *Sahel Syndrome*.

A   Assignment of the relevance criteria (R_1 - R_6) to the three selected complexes relating to the *Sahel Syndrome,* and the integration principles (I_1 - I_{12}) necessary for handling the latter.

B   Evaluation of the three complexes of questions pertaining to the *Sahel Syndrome*. The following aspects were assessed: the degree of horizontal integration necessary for handling the respective complex and the significance of the latter for solving problems related to the *Sahel Syndrome* (vertical integration). The values indicate the high priority of the three complexes for research on the *Sahel Syndrome*. Assessment was on a scale between 1 (very low) and 5 (very high).

Source: WBGU

		Issues relevant to the *Sahel Syndrome*		
		Response options of the population affected (marked red)	Regional climate change (marked green)	International economic conditions (marked blue)
A	Relevance criteria	R_1, R_2, R_3, R_5	R_1, R_2, R_4, R_5	R_1, R_3, R_4
	Integration principles	$I_1, I_2, I_4, I_7, I_8, I_9, I_{11}$	$I_1, I_4, I_5, I_6, I_8, I_9, I_{12}$	I_3, I_4, I_7, I_8
B	Horizontal integration	4	3	3
	Vertical integration	5	3	4

al and case studies and to generalize them into typical response patterns. At the same time, consideration must be given to the conditions that are created by the ecospheric or anthropospheric dispositions, the exposition factors as well as the specific manifestation of the *Sahel Syndrome*, and those which have resulted during the historical process (I_2). Work on this aspect could be linked up with the DFG priority research program Human Dimensions of Global Environmental Change (I_8). In addition, links could be forged with the future research foci of the IHDP (*see Section B 1.3*) – with activities on the *Social Dimensions of Resource Use*, for example (I_9). Many geography departments and institutes at German universities possess case-related and regionally specific knowledge. What is required is a targeted evaluation of results, generalization beyond the region in question and integration with theoretical aspects.

* *Modeling and simulating the behavior of exposed groups.* The experience gained from case and regional studies must be processed by systems analysis and the findings on human behavior and response strategies applied to the specific *Sahel Syndrome* context. Possible methodologies include the use of modeling and simulation techniques. For instance, one could generate rules for simulating migration motives and flows in order to estimate food scarcity and refugee flows. In this connection it would make sense to enhance collabora-

tion (I_7, I_8) with environmental psychology and sociology, whose research has focused predominantly on response options and action space restrictions in the industrialized nations. Greater use could be made of modeling and simulation techniques (I_4), for example, as is the case with the research on human behavior in complex non-transparent situations conducted by the psychology department of the University of Bamberg. The results of social dilemma research, as supported by the DFG, should be evaluated and specified with special reference to the Sahel Syndrome (I_8).

* *Elaboration of response alternatives based on experience and models.* This is of great importance for developing possible solutions (*vertical integration*), in that attention is focused on dilemma situations typical for this syndrome. The integration principle of relevance here is participation (I_{11}). Experience gained through development cooperation must be put to use here and the existing knowledge potential in Germany determined. In Germany, the basis for this could be a temporary research network (I_7) operated by a "Blue List" economic research institute such as the German Institute for Development Policy (DIE, Berlin), universities (e.g. development-oriented women's studies) and agricultural research institutes. Furthermore, research on *empowerment* (AGENDA 21) could also be included in order to enhance the position of marginal groups.

6.3.2
Complex 2: Regional Climate Change

In most cases, marginal agricultural regions are especially vulnerable to changes in regional climate. Depending on the nature and intensity of these changes, greater soil degradation and/or freshwater scarcity are the probable outcome. In addition to natural climate variability, a special focus of interest concerns anthropogenic causes (which can be influenced directly) – on the one hand, the enhanced greenhouse effect primarily induced by the industrialized countries (R_4), and its specific manifestations at local level, and, on the other, the regional climate changes resulting from local human activities. The latter are predominantly a consequence of land-use changes, which may lead to changes in evapotranspiration, albedo and surface roughness and which are closely related to anthropospheric aspects of the *Sahel Syndrome*. Declining agricultural yields possibly caused by regional climate change then lead to further land-use changes, which in turn have a feedback effect on regional climate.

Some features of the *Sahel Syndrome*, such as shifting cultivation, contribute to the global greenhouse effect on account of the CO_2 emissions produced by slash-and-burn agriculture, so that a feedback effect – albeit weak – exists here, too. Because of the stronger feedback loops, however, the regional dynamics of climate change cannot be investigated unless due regard is paid to the bases of life of the people affected and their possible responses. These interrelations have been little explored to date and are accordingly little understood (R_3), which makes it all the more important to improve such understanding in order to determine the impacts of the greenhouse effect on the food supply of a significant portion of the world population (R_1).

Although global predictions of climate changes resulting from the greenhouse effect have now become feasible within the German climate research program (R_6), e.g. through the establishment of the German Climate Computing Center in Hamburg (DKRZ), methods have yet to be developed for transposing these global results to the Sahel-disposed regions. The *downscaling* rules required to derive the most important variables of local weather have to take into consideration the land-use changes that are relevant for the regional climate and which in some cases have occurred in response to climate changes. An interdisciplinary institute like the Potsdam Institute for Climate Impact Research – cooperating, for example, with the university institutes actively involved in the IGBP-BAHC (Biospheric Aspects of the Hydrological Cycle) and EFEDA (Echi-

val Field Experiment in a Desertification Threatened Area) research programs – is capable of handling this intricate network of problems (I_6) by examining meteorological aspects concurrently with agricultural and ecological issues. Mathematical modeling will certainly play a major role in this context (I_4).

A specially tailored research program (I_8) applying mainly socioeconomic models is needed to investigate the land-use changes that occur or are likely to occur in regions with a disposition towards the *Sahel Syndrome* in response to possible production losses.

Remote-sensing time series of land-use changes, such as those carried out in the Humid Tropical Forest project (USA) for tropical rainforests, could serve as a important empirical foundations for such modeling (I_5). Coordination of such activities has just begun within IGBP/IHDP-LUCC (Land-Use and Land-Cover Change). The German research community has an opportunity here to strengthen its international orientation (I_9) through appropriate contributions. Moreover, an expert evaluation of activities can help ensure that research is conducted across disciplinary boundaries (I_{12}). The emphasis within such evaluation would be more on the structure rather than the "value" of research, and the process would have to be concentrated on feeding results back to the relevant research groups; all this calls for new methods of research evaluation.

An example of existing expertise on the research issue described here is Section D of Collaborative Research Center 268 (Geography of the West African Savannas) and Collaborative Research Center 69 on Problems of Arid Regions from the Earth Science Perspective (R_6). WAVES (Water Availability, Vulnerability of Ecosystems and Society) is a bilateral research project between Germany and Brazil which attempts to integrate the socioeconomic and ecospheric analysis of the Sahel-prone northeast region of Brazil (I_1).

The disposition map in *Fig. 11* is a useful starting point when planning further research projects that are integrated through a common reference region (I_1). In those regions with a medium to high disposition to the Sahel Syndrome in which the syndrome has not yet emerged with any intensity, investigating the close links between regional climate change, trends in agricultural yields and land-use changes may be of crucial importance in preventing the further spread of the syndrome.

6.3.3
Complex 3: International Economic Conditions

Both national and international economic trends or frameworks (e.g. the globalization of markets, international indebtedness, the world trade regime) may operate as major causes of marginalization within the *Sahel Syndrome*. In order to avoid the vicious circle peculiar to the syndrome, it is essential to know the role played by economic trends and conditions in social marginalization processes, and the conditions that must be present before there can be any avoidance of or adaptation to the specific problems. Given its role in the world economy, Germany has both a special responsibility (R_4) as well as an important role to play in defining and giving shape to this issue in an appropriate research program.

The main features of non-adapted national economic policy within the *Sahel Syndrome* are that it

- is geared too much to securing an adequate livelihood for the urban population, and not enough to the problems faced by agricultural producers;
- relies too heavily on export-oriented monocultures while neglecting to ensure food security through the development of a local agricultural sector;
- prevents sustainable forms of land management by creating the wrong incentives.

Many of these factors are reinforced by international influences: agricultural development is blocked because of imports from countries with highly subsidized agriculture, high indebtedness induces a short-term orientation, while credits are linked to certain development paradigms and the related criteria imposed by international institutions (e.g. structural adjustment programs).

The research necessary in this connection can be categorized as follows:

- *Studies on the determinants of agricultural production in the developing countries.* Because of its disciplinary orientation, agricultural economic research often runs the risk of neglecting social, cultural or ethnic factors, despite the fact that these are generally of enormous importance for regional economic development (I_3). Existing research approaches differ, moreover, in the degree to which they take account of the world economy and its structures. An integrative approach is essential here – one example to follow is provided by the Institute for Agricultural and Social Economics in the Tropics and Subtropics at the University of Hohenheim, where four departments are collaborating (I_6).
- *Analysis of global economic structures as a prerequisite for improving socioeconomic conditions in developing countries.* The disposition space of the *Sahel Syndrome* shows that most of the regions potentially or currently afflicted are located in developing nations, newly-industrializing countries or economies in transition. In some cases, the degradation processes are influenced to a major degree by global economic structures. The prevailing segmentation of economic and social research on these issues until now means that greater integration is needed as a matter of urgency. Temporary networks (I_7) as well as appropriate support structures (I_8) for joint projects involving the latter two research fields could trigger new impulses for German global change research. It would be conceivable, for example, to set up cooperation between one of the economic research institutes, the German Institute for Development Policy (DIE, Berlin) and the research institute of the German Foreign Policy Association (Bonn). Possible networks or programs may evolve, *inter alia*, from institutional economics or evolutionary economics, which study (world) economic mechanisms with reference to social and cultural institutions.

- *Empirically based regional research.* Case studies on the socioeconomic and sociocultural conditions governing the outbreak and trajectory of the Sahel Syndrome already exist for a considerable number of countries where the "illness" has already broken out. Numerous institutes at German universities could be mentioned in this connection, particularly in the fields of geography, agricultural economics and the sociology of development. Future-oriented global environmental research can build on the work already accomplished in these studies. The next step, however, is to overcome the boundaries between disciplines when investigating a specific region and to integrate the analysis of transformation processes at regional and global level.

6.4
Organizational Implications

The intricate links between the various questions raised above necessitates careful coordination of research activities on the main complexes. For this reason, the Council deems it urgently necessary to set up a research network under a joint management body in order to ensure adequate coordination of work in the various programs and institutes. The basic structure of the Sahel research network to be set up is depicted in *Fig. 12*. In addition to hierarchical coordination (*Fig. 8*, black arrows), special attention must be devoted to consistency in the design of analytical and methodological integration principles. The modeling and simulation methods used to assess the role of international economic structures, for instance, should be equally compatible to those with which the range of options open to the marginalized population is studied and to those used in regional climate modeling. The same applies to the other principles (socio-ocultural structures and processes, spatial reference, etc.). Establishment of a research network is essential to achieve these aims, whereby the GTZ, for example, could play a role in the joint management body. The network's structure should be aligned with the aforementioned links between the respective research questions, portrayed in *Fig. 12* as steps 1-4. The links generated by applying the integration principles take the form of the connecting networks in step 4.

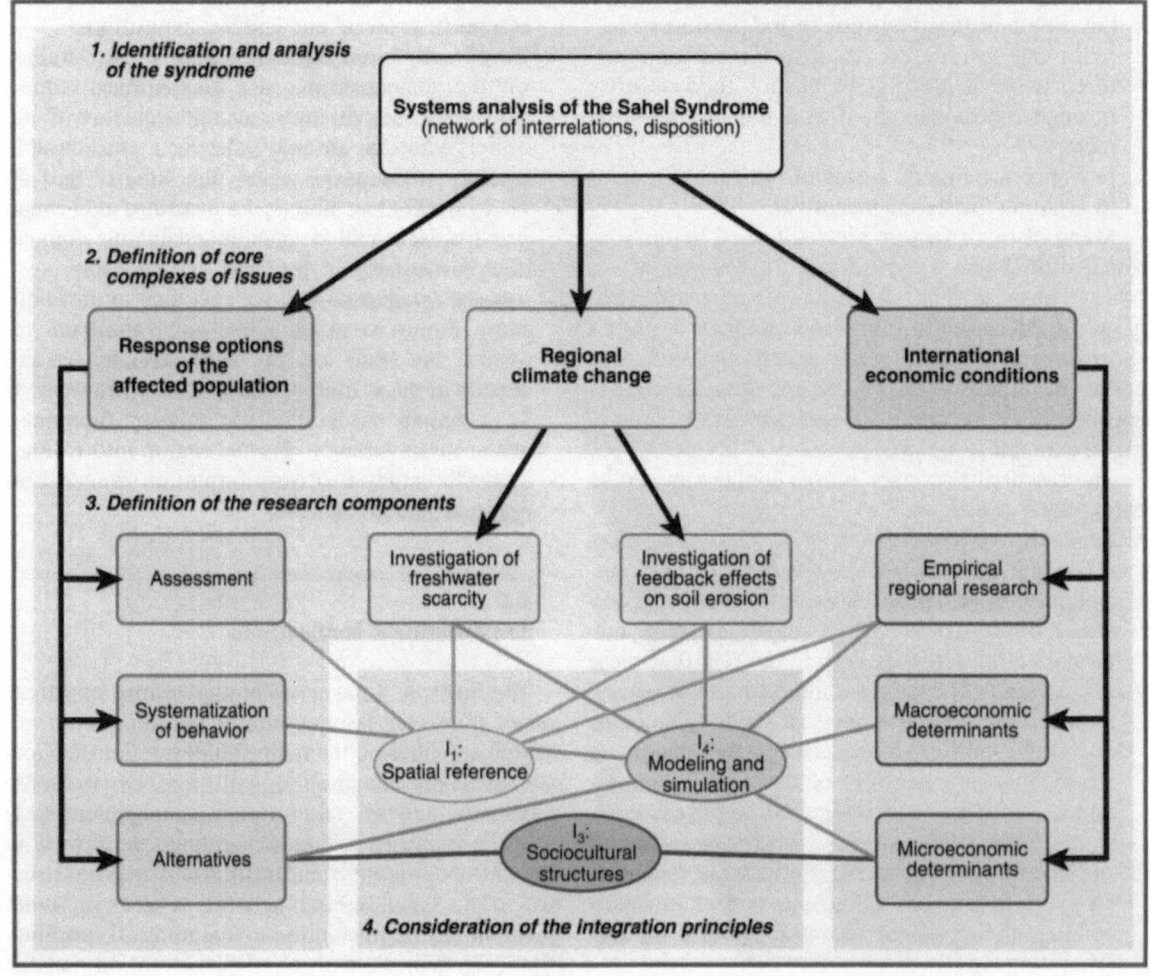

Figure 12
Basic structure of the research network for analyzing key issues relating to the *Sahel Syndrome*. Parts 1 - 4 stand for the methodological sequence in elaborating the Sahel research network. The black arrows represent hierarchical coordination within the research network. The colors of the complexes correspond to those in *Fig. 8*. The colored connecting lines in the fourth field stand for the integration principles that are observed when developing the Sahel research network and which have to be applied uniformly and generally across all the individual research issues.
Source: WBGU

7.1
Special Features of the Problem-Solving Process

Section B of the Report describes the status of and gaps in the research on global environmental problems in specific compartments of the ecosphere and anthroposphere. In addition to these specific research issues concerning concrete manifestations of global change problems, it is also necessary to examine the problem-solving process itself.

Research on decision-making processes in the field of environmental policymaking has mainly been concerned with problems of national environmental policy. Although this has led to results that also have a bearing on the environmental decision-making process in the international and global framework, the situation for policymaking is actually more complex. Global problems tend to be long-term in nature, which gives rise to major problems with regard to diagnosis and prediction. This results in special demands on early-warning systems and planning instruments, as well as on research methods and instruments. Global problems are also much more complex than environmental problems at the purely national level, with all the implications for consensus-formation and choice of instruments that this entails. Conflicts over the ends to be pursued are more difficult to resolve in an international context due to differences in culture, religion and level of development, the latter especially.

Research methods and approaches geared to national environmental policy must therefore be designed in such a way that they can also be applied to the constituent elements of the decision-making process relating to global environmental change. The focus should not be restricted to specific disciplines. The overriding issue is to structure the various elements constituting the problem-solving process, and then to ask which disciplines have already contributed and/or which disciplines should make a greater contribution to interdisciplinary research in future.

The next step is to ascertain which results are already available and in what respects they require

supplementing. A distinction can be drawn between the following elements of problem-solving processes:
- *Initial treatment of the problem.* The starting point for solving global change problems is analysis, i.e. the identification of causes and effects and the assessment of future trends (forecasts). Due to the complexity of the material and the need for integrated research approaches, it is necessary to have an adequate methodology, such as *systems analysis* (*Section C 7.2*), for describing and explaining the problem and arriving at forecasts.
- *Guiding principles and objectives.* Once the problem has been analyzed, it is necessary to define guiding principles and objectives. In the view of the Council, there are serious gaps in *research into guiding principles*, which needs to be oriented towards the concept of sustainable development and made more specific by means of relevant principles for action and appropriate indicators (*Section C 7.3*).
- *Responsible bodies.* Policies for influencing global environmental change require appropriate agencies at various levels (global, regional, national and local). Because sovereign states are the bodies which take action at international level, special attention must be focused on the mechanisms of decision-making and action which take place there. The specific constellation of responsible bodies and the problem of how to achieve effective cooperation between them must therefore be studied in greater detail. Suitable methods for such research have to be selected and/or refined, *game theory* being one example in this context (*Section C 7.4*).
- *Instruments.* Agreement on objectives is achieved using instruments which are either already available to global environmental policymaking or which have yet to be developed. Research into the strength and effectiveness of these instruments is needed and must be advanced further. In particular, research is needed on transsectoral instruments, such as the various international environmental conventions, but also on the various subor-

dinate instruments which operate under those conventions (*Section C 7.5*).

- *Implementation*. Once international conventions have been adopted, the next question concerns their *implementation* and *enforcement* as well as options for imposing sanctions. In view of the fact that problem-solving processes often stagnate at precisely this stage, the obstacles which arise in this context must be subjected to precise analysis (*Section C 7.6*).

- *Research into decision-making and risk management processes*. In addition to research into the above elements of the decision-making process, especially the problem of responsible bodies and the effectiveness of policy instruments, it is essential to conduct research into decision-making and risk management, which involves investigating two specific features of the problem-solving process, namely the problem of consensus-formation in cases where there are fundamental disparities of interest, and the question as to how to handle uncertainty (*Section C 7.7*).

The rest of this section analyzes existing research on the above steps and identifies areas where there are still major gaps. The question as to which institutions are capable of conducting this research will not be dealt with at this point, but is treated elsewhere in the Report. In *Section C 8* the Council proposes, for example, a German Strategy Center on Global Change, whose functions would include interdisciplinary tasks of this kind.

7.2
Decision-Oriented Problem Analysis

This area embraces all research on the ecosphere and anthroposphere which is aimed at identifying, explaining and predicting global environmental problems. In addition to the basic analysis of specific environmental problems, research must meet the following requirements:

- Different *perspectives* have to be taken into account when describing and explaining the problem (different disciplines, actors, perceptions of the problem and explanatory approaches, etc.); this necessitates discourse between the scientific community, policymakers and the general public.

- Appropriate *information systems* must be created to strengthen links between the sectors and domains.

- These systems must be designed to provide *early warning*.

- Environmental policy decisions relating to global change are generally made under *conditions of uncertainty*. The relevant decision-making processes

are analyzed from the perspectives of various disciplines within the framework of *decision analysis* and *risk management* (*see also Section C 7.7*).

The complexity of global problems means that priority must be attached to developing and refining methods that apply an integrative approach to assess causes and effects (e.g. the Syndrome Approach, *see Section C 2*). Systems analysis and its research tools, namely the modeling and simulation of complex systems, must be advanced as a matter of priority.

7.3
Developing Guiding Principles and Research on Goals

The international community has been committed to the guiding and binding principle of *sustainable development* since the 1992 UNCED conference in Rio. Applying the principle involves integrating and aligning all national environmental policies within the framework implied by the concept (WBGU, 1993; SRU, 1994). Defined goals and the procedures by which they are elaborated must therefore be oriented to the concept of sustainable development, which requires a systems analysis of ecological, economic and sociocultural elements. The Commission on Sustainable Development (CSD) is responsible for specifying the concept, in particular by defining indicators of sustainable development, and for monitoring compliance with the pledges made at the Rio Conference.

There are several initiatives in the area of national environmental policy that can be drawn on when applying the sustainability concept to specify objectives. Both the SRU's environmental report (1994) and the report prepared by the Enquete Commission Protection of People and the Environment (1995b) show possible ways to implement and quantify sustainable development targets (and the shortcomings that exist in this connection).

Proceeding from a systems analysis and the realization that solutions must take the anthroposphere as their starting point, the Council regards the following as key goals for global environmental policy:

- Social development must be understood as more than rising material affluence. The food and health situation as well as educational opportunities have to be radically improved on a global scale.

- A step-by-step equalization of material and immaterial living conditions between highly developed and less developed nations must be striven for (*intragenerational equity*).

- A special challenge for research concerns the interests of future generations and ensuring that these are adequately respected (*intergenerational equity*).

Elaboration of such a guiding principle is a *discursive process* that remains open to adjustments over time. The concept of sustainable development cannot be defined as a fixed variable for all time because, as an expression of social values, it is subject to change. Those guiding principles that ultimately represent a code of appropriate environmental conduct require a broad *discussion* at all levels of society, not only to ensure adequate acceptance of the necessary changes in behavior, but also to exploit the creative potential of participative processes.

Despite intensive discussion, the concept of sustainable, viable (WBGU, 1993), lasting and environmentally sound (SRU, 1994) development has not yet been specified in sufficient detail. No consensus has been reached, either in Germany or in the EU, and especially not at global level, on how to operationalize this guiding principle, implementation of which requires a wide variety of competencies (SRU, 1994 and 1996). For this reason, research must be conducted both on the barriers that are preventing specification of this system of goals, and on the options available for overcoming these barriers, e.g. environmentally compatible lifestyles.

The Council's approach in this connection is illustrated by the *"crash barrier model"* (*Section C 2*), designed as an alternative to some finite "limiting state". The crash barrier model builds on a policy approach that specifies limits but permits free decisions within these limits. One of the tasks of the research community is to clarify the extent to which the guiding principles and goals of specific environmental conventions conform to this principle (*Section B 3.7*). Another task is to adjust the model to additional boundary conditions generated by interdisciplinary research and to integrate ecological, economic and social aspects.

In the view of the Council, the lack of consideration and specification of the needs of *future generations* is a great shortcoming in research on goal-finding. International and interdisciplinary research must be conducted on the assessment of future needs, whereby ethical aspects must also be taken into account (e.g. legitimation for discounting the benefits accruing to coming generations) (*see also Section B 3.6.4.1*).

Another topic meriting detailed analysis is the extent to which the *materials flow management* approach proposed by the Enquete Commission of the German Bundestag could operate as a new regulatory instrument, or whether it would require further rules instead. The goal, or demand, that ecosystems not be stressed beyond their absorptive or carrying capacity raises the question as to the precise limits to

be observed according to the precautionary principle, or the dictates of intergenerational equity. Research is needed here, whereby special attention must be devoted to potentially irreversible changes.

7.4
Research on Global Environmental Institutions

Whereas other elements of the decision-making process involve gradually increasing complexity in the transition from national to global environmental problems, the situation with intergovernmental institutions features a qualitatively different dimension (Zimmermann, 1992). For instance, there is no central authority with power to enforce global environmental policy. Usually, one would conceive such a supranational institution as something similar to UN institutions. The latter's main function is coordination, however, rather than issuing directives and imposing sanctions for non-compliance. Global environmental policymaking bodies at national level are not subordinate to a sovereign "world government", so they can assert their individual interests at any time. The particular degree of involvement, whether as a causal agent or as a victim of global change, must therefore be integrated in strategic analysis, as must the special role of politicians. A promising set of analytical instruments for examining interdependent decision-making situations involving partly conflicting interests and possibilities for cooperation is provided by *game theory*.

Another major research issue concerns the interests of actors within the framework of existing or planned conventions. To what extent are these interests shaped by the level of development, societal values, and religious or ethnic affiliations? Which actors are receptive to the idea of global responsibility? What are the steps to jointly agreed solutions to global change problems, when conditions for the parties involved differ so greatly?

7.5
Research on Instruments of Global Environmental Policy

The characteristic difference between the instruments of national and global environmental policy (*see Section B 3.6.4.1*) has something to do with the special qualitative features of the policymaking bodies in each case – there being no such thing as a "world government", instruments at international level must be relatively "soft". Enforcement of environmental policy by and within nation-states is a different matter – here it is possible to deploy familiar instruments such as command and control, levies, liability law, and environmental education programs. The debate over which instruments will best ensure the effectiveness of global change policy has been confined to the traditional dichotomy between regulatory and market-based instruments, and must now be widened to include educational and psychological approaches and strategies for modifying behavior (in the form of a broadly conceived *environmental education, see* WBGU, 1996) as additional instruments of environmental policy.

Other kinds of international instruments need to be examined to determine how useful they might be in achieving the goals of global environmental policy. Options here include declarations, "round tables", or national measures such as certification (e.g. the "Carpets not produced by children" label). Research is also needed on how effectively the world economic order can be "greened" by restructuring international liability law, operationalizing the concept of "eco-dumping", implementing international certificate models, etc. Studies should also be conducted to find out the extent to which informational instruments, and above all the EU framework for environmental management and audits (WBGU, 1996) can be deployed in response to global problems.

It is also necessary to extend interdisciplinary research to global conventions (*see Section B 3.7.2.1*). Although extensive research has already been conducted into environmental conventions with respect to creation, agenda setting and regime formation, problems related to regime implementation and regime effectiveness have only been touched on thus far. What is needed are links between research in political science and law, on the one hand, and the theory of political economy, on the other. Such interdisciplinary research should then focus on the following aspects:

- *Regime formation.* A number of studies have already been carried out on the "law of the sea", "ozone" and "climate" regimes. In the future it will be important to systematically retrace the negotiation processes leading up to the Climate, Desertification and Biodiversity Conventions, and to examine those involved in planned conventions (forests, soils, water).
- *Agenda setting.* The process of agenda setting is a basic prerequisite for the formation of environmental regimes, but has been the subject of little research to date. The central question is how issues are put on the international policy agenda – i.e. who defines and justifies particular issues and ensures they have priority over competing issues.
- *Regime effectiveness, regime implementation.* In addition to research on the preconditions for environmental regimes, it is necessary to gain greater

insight into regime effectiveness. Methods for defining and measuring effectiveness have to be developed, particularly in the field of empirical political science.

7.6
Research on the Implementation of International Conventions

Once a global convention or protocol has been adopted, a number of tasks must be performed to ensure successful *implementation.* These include:
- Interpretation of the basic principles underlying the convention.
- Specification of the agreed objectives.
- Discussion and assessment of the instruments to be applied.
- Monitoring the agreed mechanisms for coordination and consultation.
- Verification of national contributions.
- Definition and enforcement of sanctions.

Supporting research is needed for effective performance of these tasks. Research results are still rather sparse in this area.

7.7
Decision Analysis and Risk Research

Finding solutions to the problems associated with global change is an extremely difficult process for two reasons. Firstly, the parties involved in this problem-solving process are sovereign states with highly divergent interests in many cases. Secondly, there is often a lack of certain knowledge on which to base decisions (see the references to the uncertainty of the knowledge base in the most recent IPCC report, 1996).

These hindrances for the problem-solving process are examined more closely by *decision analysis* and *risk research.* The latter, for example, examines the perception, assessment and acceptance of risks. In the context of the problem-solving process, therefore, it plays an important role in identifying environmental problems. Moreover, the selection of instruments and responsible bodies depends on the respective assessment of the risks. In the view of the Council, this complex of problems still requires considerable researching, involving not only explicit risk research, but also dilemma and values research, research into behavioral determinants, control and intervention systems, the effectiveness of media and communication strategies, and environmental assessment methods.

One can distinguish in the field of *decision analysis* between two types of approach. Firstly, that taken by social and behavioral science, involving research on the effects of mediation processes, for example, or on "decision-making under conditions of uncertainty" (*see Section B 3.8.2.3 and B 3.8.4.1*). Secondly, there is the normative and rather formalist theory of decision-making developed primarily by economists. The two approaches should be examined with regard to the applicability of their methodologies and procedures to decision-making on global problems. The results of this analysis should be used to refine the current approaches, whereby greater collaboration between experts in economics, social science, behavioral science and ethics should be striven for.

8.1
Requirements

Environmental research addresses specific problems and tends to be synthetic rather than analytical. This applies all the more to global change research and its focus on the Earth System as a whole, so the question is raised as to the principles by which such a problem-oriented synthesis can be achieved. Repeated calls for "networking", "interdisciplinary thinking" and "cooperation" are certainly not sufficient – what is really needed are organizational principles and instruments that enable a holistic analysis of global change syndromes.

Several requirements can be formulated for the future organization of global change research in Germany:

- New ways of *defining research issues*. The phenomena of global change involve complex problems that have to be analyzed through interdisciplinary collaboration, and split up into components of manageable proportions for projects. One way this can be done is by applying a framework based on the Council's syndrome concept.
- New ways of *performing research*. The demand for more inter- or transdisciplinary research suggests that researchers work more in groups and that regional and transregional research units as well as *inter-institutional research* networks be formed on a temporary basis.
- New forms of *research review*. The current review system must be subjected to critical examination to identify gaps and structural faults regarding its suitability for handling complex, transdisciplinary issues. Assessment by a team of experts is recommended, similar to the procedure for collaborative research centers.
- Adjustment of *research funding*. New research approaches (especially transdisciplinary research and research on issues with an extended spatial and/or temporal frame, e.g. longitudinal studies) necessitate adjustments of financial provision if they are to produce the additional output expected of them. Funds must be provided for an infrastructure that promotes interdisciplinarity and for stimulating transdisciplinary "translation processes". Given current financial constraints, "new" funds can only be part of the answer. What is important, therefore, is flexibility in the deployment of existing project funds.

Major premises for the discussion of such new demands regarding the funding and organization of global change research are as follows:

- Global change research focuses on interrelations between the ecosphere and the anthroposphere and thus involves most of the scientific disciplines. It is very unevenly developed across the various disciplines, however. While already well advanced in the natural sciences, there is much catching up to do in the social and behavioral sciences, economic and political science and law (*see Sections B 3.6 - 3.8*).
- For various reasons, global change research has lacked *problem-solving competence*. Since solutions to problems always originate in the anthroposphere, research geared to the development of solutions will fail in this endeavor unless it integrates the outcomes of human science research. Solutions involving the actions of individual people are primarily of a local nature, so that a division between global change research and "other" kinds of environmental research is not (always) useful for research geared to intervention.
- On the one hand, the scientific community must find out what decisions have to be made on the part of policymakers, administration, industry, etc. with regard to global environmental problems. It must translate the respective needs into research issues that are dealt with on a multi- or interdisciplinary and competitive basis (competition between concepts, methodological approaches, etc.). On the other hand, the results of global change research must be "translated", processed and communicated in a form that is relevant for decision-making so that they have a chance of becoming relevant for policymaking and action.

8.2
From Multidisciplinary to Transdisciplinary Research

Research projects and programs continue to be designed, funded and performed predominantly along "traditional" lines of thinking, with discrete disciplines, conceptions of the world and methodologies serving as orientation. Even in programs with explicit interdisciplinary design, discourse across the boundaries separating disciplines only occurs alongside project work within the disciplines. In effect, no genuinely *inter-* or *transdisciplinary* perspective – from the definition and description of problems, through the design and implementation of research projects to user-oriented translation and communication of results – has yet emerged.

There are many *reasons* for this disciplinary orientation: the structure of university research and teaching evolved for decades on the basis of specific disciplines, for example, while research funding reflects this disciplinary structure. The latter is perpetuated in Germany by the appointments procedure, and by the resultant lack of career opportunities for researchers with an interdisciplinary approach. The high degree of specialization that has been achieved in most disciplines promotes the formation of smaller and smaller "niches", which are often "occupied" by only a few researchers or research groups whose theories, topics and results appear less and less communicable to representatives of other disciplines or even to those within their own subject area. Another hurdle for inter- or transdisciplinary global change research is the heterogeneity of disciplines with respect to their theories, concepts, basic premises, conceptions of humanity and methodologies. Even if all external obstacles were to be removed, the individual researchers themselves would still have to invest the time and interest needed to grapple with and assimilate "extradisciplinary" theories, issues and results before any interdisciplinary work would be possible.

It is hardly surprising, therefore, that the interdisciplinarity necessary for global change research remains a somewhat abstract construct as long as the reality is at best a *multidisciplinary* "patchwork" that mostly exhibits only loose links between research activities in the diverse disciplines.

Environmental problems "frequently fail to do us the favor of being defined as problems for disciplinary specialists" (Mittelstraß, 1989). Indeed, the first task is to even define these problems in the first place – a task that is rendered difficult by the different perspectives of the various disciplines, but which is absolutely indispensable for a problem-solving orientation in research. The term *transdisciplinarity* was coined for this multi-perspective procedure that commences with the definition of the problem and must lead from there to modified research practice. Transdisciplinarity removes the territorial boundaries between disciplines, thus permitting a new perspective on the problems. However, this does not make disciplinary research obsolete – on the contrary. The transdisciplinary perspective depends on the existence of specialized disciplinary perspectives, but compels researchers to develop integrative concepts and to qualify individual standpoints in the light of others.

To conduct global change research, it is necessary to look for suitable methods, criteria and guiding principles for integrating and processing the research results already generated within the separate disciplines. Secondly, strategic interfaces between the individual disciplines need to be identified – not only between the natural and social sciences, but also within these sciences. Once these steps have been taken, it will be possible to define transdisciplinary research needs. A possible starting point for this process is the *syndrome approach* developed by the Council.

In addition to the demand for synthesis and holistic approach, it is essential not to lose sight of the fact that global change in the ecosphere and in the anthroposphere is based on phenomena which can only be studied through systematic, knowledge-generating, disciplinary research. While it is true that politicians cannot wait to make decisions until there is scientific certainty about the causes of certain global change phenomena, the quality of directives and predictions depends on how much knowledge is actually available about the processes that lead to global change in the ecosphere and the anthroposphere. For example, had it not been for the basic research work on stratospheric chemistry in the 1970s and 1980s (awarded the Nobel Prize in 1995), and the many years of routine monitoring of ozone concentrations over the Antarctic, there would never have been a Montreal Protocol on Substances that Deplete the Ozone Layer.

8.3
Organizational Implications

An appropriate organizational framework must be created for the identified fields of global change research. This calls for review and improvement of existing principles and instruments of research organization. New instruments must be tested and established wherever the problem-solving orientation of global change research calls for them.

Generally speaking, Germany's research landscape provides a good basis for synthetic as well as analytical global change research, thanks to the wide range of institute types (those of the Max Planck Society, the "Blue List", the Fraunhofer Society and the university institutes, as well as large-scale research institutes and those funded by federal and state agencies), their varying, but overlapping spheres of activity and the fact that they are usually well-staffed and well-equipped. Germany holds a leading position in the world in certain fields of climate, atmospheric, marine and polar research. However, there is a recent and disquieting trend towards non-selective staff cuts, which deprives research institutions of the flexibility to explore new lines of research and opportunities for collaboration. There is much catching up to be done in the analysis of people-environment relations outside the natural sciences, because many of the human sciences have been slow to address the environmental implications of individual and social action. In future, however, it will be crucially important that the natural and human sciences collaborate in genuinely transdisciplinary research, as the Council has explained throughout this Report.

Funding of global change research must ensure, on the one hand, that there is reliable planning of national and international research projects over several years and, on the other, that there is sufficient flexibility with respect to focal points and methodologies. The differing time scales in global change research activities (e.g. longitudinal studies in environmental and social scientific research) must be taken into account when assessing the time frames for research funding. Moreover, continuous operation of environmental and also social monitoring programs is essential for a large proportion of global change research. Scientists interested in global change issues must be offered incentives to make joint applications for research funds within the framework of priority programs, collaborative research centers and network projects.

8.3.1
Make Better Use of Existing Instruments

In recent decades, the DFG and the BMBF have developed a number of funding instruments which meet the needs of global change research and should be utilized by the latter to an increasing extent in future. Of special importance in this context is the funding of projects that are applied for jointly by researchers from different disciplines or "spheres"; examples include the priority programs and collaborative research centers of the DFG, or the network projects of the BMBF. The report of the *Wissenschaftsrat*

(German Science Council) on environmental research in Germany (1994) contains an overview of the priority programs and collaborative research centers of the DFG in the field of environmental research; most of them are closely related to global change.

COLLABORATIVE RESEARCH CENTERS

The *collaborative research centers* of the DFG enable treatment of a certain topic (e.g. aspects of a specific syndrome) by a large number of researchers and research groups from different fields. Since the issues of global change are especially well-suited for bridging the gap between the ecosphere and the anthroposphere, more "transboundary" collaborative research centers should be set up in this connection. A possible topic would be Decoupling of Material Flows through Land Use – a Cause of Soil Degradation and Pollution of Neighboring Systems. This phenomenon involves a complex web of causes and effects, and operates within several syndromes at local, regional and global level. This is the framework within which the natural sciences need to investigate the basic features of decoupling, the impacts this has on utilized ecosystems, as well as the effect of exported substances on neighboring systems. Because sources and sinks can be geographically separated by large distances (e.g. Dutch vegetables exported to Berlin, soybeans from Brazil to the region around Vechta in Lower Saxony or flowers from South America to Germany), it is also necessary to incorporate the socioeconomic background into the analysis of decoupling processes.

The restrictive requirement that collaborative research centers be located at individual or immediately neighboring universities, should be relaxed in view of the opportunities provided by modern telecommunications, and a fundamental boost given to inter-institutional research.

PRIORITY PROGRAMS OF THE GERMAN
RESEARCH FOUNDATION

Although the majority of the *DFG priority programs* are disciplinary in orientation, they also offer opportunities for inter- or transdisciplinary research. Colloquia, workshops, working groups and bilateral meetings enable links to be forged between the individual projects.

The DFG priority program on Human Dimensions of Global Environmental Change is a first example where this instrument of research funding has been deployed to support interdisciplinary work on global change. The main participants in the program, which commenced work in 1994, are research groups from psychology, sociology, geography, social anthropology, political and economic science departments at various German universities. However, future priority programs of this kind should also be utilized as bridging devices between the natural and the social sciences.

NETWORK PROJECTS

The network projects funded by the BMBF in the form of multiannual research programs – one example being the BMBF framework for Ecological Research in Urban Regions and Industrial Landscapes (Urban Ecology) – offer special opportunities for problem-solving research approaches. These network projects are affiliated to existing research institutions and often include several institutes. The topics "Hydrological Cycle in Urban Areas", "Environmentally Sound Mobility" and "Resolving Land-Use Competition" provide not only for collaboration between the natural and social sciences, as the titles suggest, but also for direct cooperation between research groups and practitioners from selected local governments. Two large-scale projects set up within the framework of economic cooperation with Brazil pursue a similar strategy: Brazilian and German scientists from the fields of hydrology, climatology, agriculture, economics, landscape ecology and psychology are involved in the project Water Availability, Vulnerability of Ecosystems and Society in Piauí (WAVES). Local groups were integrated into the research work from the outset, the aim being to ensure sustainable regional planning with special reference to migration problems. A similar approach is applied in the Mangrove Dynamics and Management project, where the intention is to provide the scientific basis for the protection and sustainable use of the tropical mangrove belt. Ecologists and sociologists are collaborating closely here. Another example of a transdisciplinary network project for the study of global change is the joint research program of the BMBF and the northern German *Länder* entitled Climate Change and Coastal Regions (*see Box 8*).

MULTIDISCIPLINARY INSTITUTES AND INSTITUTIONS

One way to organize long-term collaboration of researchers and research groups from different disciplines on global change issues is to integrate them into multidisciplinary institutes or institutions at local level. The type of research is then determined by the organizational and structural framework. Examples for this mode of research organization are – in the public sector – the Potsdam Institute for Climate Impact Research (PIK) and the Halle-Leipzig Environmental Research Center (UFZ). There are also a number of institutes and research groups in the non-public sector that study global change intensively, in many cases with considerable expertise.

8.3.2
Create New Instruments

The establishment of *fixed-term institutes* could advance German global change research significantly. The ecosystem centers and gene centers financed by the BMBF for terms of 10-15 years before being incorporated into the respective university in one form or another can serve as a model here. These institutions exhibit a high degree of flexibility and interdisciplinarity; it is not necessary when setting them up to allocate a large number of permanent positions for research and technical staff, which anyhow is mostly prohibited by present-day financial constraints. Network projects are "fixed-term institutes", too, in a certain sense. Two requirements have to be met before any interdisciplinary fixed-term institute can be established:

- Advanced research in a specific discipline is available as the basis for an interdisciplinary focus.
- A university or extra-university research institute must be willing to relocate its own resources and to make a long-term commitment to a specific research priority.

To be successful, the fixed-term institutes need a stable long-term framework (10-12 years). They should be integrated into large networks with non-university research institutes, or at least enter into firm cooperation agreements with individual institutions so that efficient communication structures are created and the research foci obtain a magnitude that ensures them a viable competitive position at international level. Fixed-term institutes involved in global change research must also integrate activities in relevant fields of social science.

Research networks (inter-institutional research) are long-term purpose-made alliances between independent research institutions for joint work on complex, usually interdisciplinary issues, and for further refinement of methodologies. The partner institutes are linked through steering committees, joint appointments and collectively utilized infrastructures. Networking can involve the staff of one institution being seconded to the research management of another institution in order to achieve more effective cooperation. An additional option is to allocate joint project posts from a network fund. This is one way to improve the situation of non-professional scientific staff currently plagued by insecurity.

An example for a research network is the planned Scientific Computing Network, which will be jointly managed by several research institutes. The objective of this network is the application, maintenance and improvement of advanced methods of scientific computing in the simulation of global environmental systems. Furthermore, the partners expect to derive particular benefits by pooling their computing resources (supercomputers, databases and communication systems, software pools, etc.) to meet needs.

Research networks are particularly suitable to the integrated study of environmental topics. More specifically, this instrument should be used to concentrate "distributed capacities" for researching the *Smokestack, Sahel* and *Urban Sprawl Syndromes,* which the Council classes as high-priority (*see Section C 5*).

Another field where a research network would be appropriate concerns threats to tropical coasts, where several syndromes of global change coincide. Such activities could be organized within a DFG priority program for natural and social scientific study of a selected group of tropical and subtropical coastal zones, thus complementing the Climate Change and Coastal Regions *Bund-Länder-Programm,* which focuses on the North and Baltic Seas. Links to the Global Environmental Facility (GEF) are recommended here for practical and financial reasons.

TRAINING AND INSTRUCTION

In its report on environmental research (1994), the German Science Council pointed out that not only research, but also *university education* on global changes processes and the interdependencies between humankind and the environment must be organized along transdisciplinary lines. The Advisory Council expressly supports the recommendations of the German Science Council: given the increasing specialization in individual subject areas related to global change at German universities, it is imperative to include the aspects of networking and interdisciplinarity in the organization of tertiary education, too. To achieve this, it is necessary to set up or support relevant courses, and within these to promote synthetic and interdisciplinary thinking among students and young researchers.

The Council therefore appeals above all for the creation and expansion of Graduate Colleges, as well as lecture cycles and general courses on global change topics. In particular, the relevant collaborative research centers should support such new forms of studying and teaching.

The comments already made regarding the organization of research apply to the field of university education as well, namely that the analysis of the complex processes of global change require an adequate level of specialized knowledge. Just as the environmental problems need to be treated in detail by the various subjects in the various degree courses in the natural and social sciences, the problems of global change must also be addressed in numerous lecture courses and dealt with in depth during advanced and postgraduate studies.

8.4
Bridging the Gap Between Research and its Practical Applications

Even if research activities are perfect examples of interdisciplinary collaboration, they will contribute little to the solution of global change problems unless the research results are adequately transformed into practical political action. To achieve this, it is necessary to consider not only the *production* of knowledge, but also the potential *uses* of knowledge – especially in the domains of politics, administration, industry, education and training, in other words at the interface between research and its application. In this connection, the Council welcomes the DFG's creation of transfer fields as an instrument for cooperation between research institutes and industry or other potential users.

A whole set of communication and transformation problems are generated at this interface, however, particularly when the context is global change. These problems are brought about through the growing volume and incomparability of available information and the considerable uncertainty attached to some of it, by the sheer numbers of actors with different interests, and, finally, by the lack of useful instruments for prioritizing options. Many researchers miss the application of their research results in concrete policies, while many users complain about the lack of realism and the inherent contradictions in the recommendations produced by the research community – assuming, of course, that these messages even reach their intended audience.

In the face of these shortcomings, an appropriate "institution" for managing the processes of exchange and channeling at the interface between research and its potential applications appears both useful and necessary. Possible responsibilities and fields of activity for such an institution are outlined in the following:

• Researchers in all subjects relevant to the particular problem *and* potential research users, i.e. decision-makers and actors at the various levels of society, must contribute jointly to a multidimensional, interdisciplinary analysis of the problems as well as to the formulation of appropriate research questions.

• On the "user side", the formal and specific requirements to be met by problem-solving global change research and its expected research results must be formulated.

- On the "research side" of the interface, existing knowledge on the specific aspects of global change must be formally integrated as extensively as possible. Before research results can be meaningfully employed to solve problems, "we have to know what we know".
- In general, research results once obtained must be processed and communicated on a user-oriented basis so that they can be "translated" into the language and action contexts of decision-makers in politics, administration and industry. This type of "popularization" of knowledge is urgently necessary for any kind of "problem-solving orientation" and should not be disregarded by scientific experts as impermissible simplification (*see also Section C 7*).
- Finally, such an institution should also act as a contact and communication center to which the providers and users of information can turn at any time, and where the relevant discourse on global change problems could be organized unbureaucratically between scientists and politicians, the media and the public.
- The formulation of specific tasks for the institution proposed here should be based on a sound assessment of communication processes between research and application. In addition, the work of the institution should be supported by research, in the form of regular self-checks.
- New instruments for financing the work of such an institution are conceivable, for example with funds provided by the German Environment Foundation or a "Global Change" Foundation (*see Section D 4.2*).

In response to the "interface problem" outlined here, the Council recommends the establishment of a *German Strategy Center on Global Change*, which would focus primarily on the transdisciplinary description and analysis of global change problems, and on the "translation" of research results into practical policies. Such a center should operate *bidirectionally* between research and application: on the one hand, it should take up the suggestions of knowledge users in the political and public spheres for a comprehensive analysis of problems, and turn these into research issues through its insider knowledge of the research landscape and current state of knowledge. Secondly, it should reformulate and disseminate existing knowledge in such a way that it effectively supports decision-making processes in politics, administration, industry, education and training.

This proposal incorporates and enlarges upon ideas that are currently being put into practice in the USA, Canada and England, to name only some examples. The most important elements of such a Strategy Center are temporary research groups seconded to the Center by their respective institutions for periods of two to three years. Potential candidates would be researchers from various disciplines with a high interest in interdisciplinary work.

Summary of Recommendations

The research recommended by the Advisory Council should be aimed specifically at gaining knowledge and generating solutions relating to the cause-effect mechanisms of global change. The most important global change phenomena for human societies, and those which directly affect Germany, are

- on the one hand, the changes in the biotic and abiotic environment of human beings:
  - climate change and sea-level rise,
  - perturbations of terrestrial and aquatic ecosystems,
  - changes in global absorption capacities and the exhaustion of those capacities,
  - changes in the stocks of renewable and non-renewable resources, and loss of biodiversity.
- on the other, changes within the anthroposphere itself:
  - growth of the world population, migration flows, urbanization,
  - growth-related use of land, raw materials and energy, and the emissions and waste thus induced,
  - globalization of the world economy and the general increase in transport,
  - excessive debt on the part of many states, leading to short-termism in political decision-making,
  - insufficient provision of financial and development aid by most industrialized nations,
  - lack of environmentalist approach on the part of international institutions (WTO, World Bank, etc.) and international regimes,
  - differences in the structure and scope of environmental legislation, and in the respective administrative practices in individual countries,
  - widening disparities in wealth between geographical regions,
  - divergent growth of environmental awareness,
  - changes in the networks for the global distribution, reactivation and expansion of knowledge.

Reference was made in the Introduction to this Annual Report to the four main shortcomings of global change research in Germany in respect of:
- international integration,
- interdisciplinary links,
- organizational networking and
- communication between the research community, the political sphere and society as a whole.

The recommendations are directed, on the one hand, at *strengthening global change research in selected fields*, and, on the other, at the *necessary development of transdisciplinary, integrated (e.g. syndrome-based) research*. These are supplemented by further recommendations regarding the organization and coordination of global change research in Germany, whereby the goal is to integrate research and environmental policy measures in a new and responsible way.

Priority Tasks in the Various Sectors of Global Change Research

2.1
Climate and Atmosphere Research

All climate and atmosphere research topics relate to global change in some way. The advanced level of German research in this field must be maintained through continuous improvement of the existing infrastructure. German climate research, for example, occupies a leading position in the world in those sectors geared to the development of coupled ocean-ice-atmosphere models, thanks to consistent support from the BMBF, the Max Planck Society and the German Research Foundation (DFG). This position can only be maintained through adequate human resources policy, continuous modernization of computing capacities and constant refinement of models. Research tasks of special relevance to global change are:

- Further development of coupled ocean-ice-atmosphere models for predicting climate along different spatial and temporal scales, and of integrated models for climate impact research.
- Research into the Earth's paleoclimate using ice cores and marine and limnetic sediments. There is a general lack of data from tropical regions and the Southern Hemisphere in this field.
- Continuation and/or commencement of measurements of the composition of the atmosphere (various characterizing substances) at selected stations in Germany and northern Europe (stratosphere monitoring), as well as at sea and in the tropics (troposphere monitoring) in the context of international monitoring programs. The German Weather Service should cooperate more closely with universities and extra-university research establishments.
- Systematic analysis of existing data from different parts of the atmosphere in order to improve our understanding of climate variability (CLIVAR).
- Development and evaluation of satellite experiments for measuring parameters and trace gases of relevance for the climate.

- Investigating the influence of aerosols and clouds on climate.
- Experimental studies of tropospheric chemistry at low latitudes (using research aircraft).

Climate and atmosphere research, in the narrower sense, is conducted primarily by natural science. Research on the impacts of global change (especially climate impact research) should involve the human sciences as well. What is needed is:

- greater development of Integrated Regional Models (RIM) and
- organization of transdisciplinary and transinstitutional research networks for studying issues of sectoral and political relevance.

2.2
Hydrosphere Research

Most topics in the field of *marine and polar research* relate in some way to global change. As with climate research, it is essential that the advanced level of German research in this field be maintained through continuous improvement of the existing infrastructure. A committed role in the Joint Global Ocean Flux Study (JGOFS), the Global Ocean Ecosystem Dynamics Programme (GLOBEC) and the Land-Ocean Interactions in the Coastal Zone Programme (LOICZ), all core projects of the International Geosphere Biosphere Programme (IGBP), is essential. Research tasks of special relevance to global change are:

- Development of the scientific foundations for an operational Global Ocean Observing System (GOOS).
- Research into human-induced impacts on marginal seas and coastal zones, and the development of the scientific basis for integrated management of coastal regions.
- Research into the polar oceans, with special reference to climatological aspects.

With regard to the *global aspects of water resources*, there is an enormous need for research into the causal interlinkages between climate, vegetation and anthroposphere, and for the development of environmentally sound land management practices that safeguard water resources in the long term, as envisaged by the Land-Use and Land-Cover Change (LUCC) and Biospheric Aspects of the Hydrological Cycle (BAHC) core projects of IGBP.

Freshwater is a vital resource in all areas of life and society, functioning as a nutrient, a cultural asset and a production factor simultaneously. The Advisory Council considers it extremely important that research into freshwater resources be intensified. Research tasks of special relevance to global change are:

- Research into the conditions for increasing the supply of water for a growing world population.
- Research into the conditions for thrifty and sustainable use of water resources, in the sense of careful management of water resources in the various sectors of use (agriculture, industry, private households) and the equitable sharing of available water (*intra- and intergenerational equity*).
- Research into the conditions for preventing the pollution of surface water and groundwater stocks.

The main focus here is to develop dynamic models of the regional and global water balance, including feedback effects to the climate system, the biosphere and the anthroposphere.

2.3
Soil Research

Soil research focuses primarily on the local and regional level, but it must now integrate global changes in climate, water balance and land use. The following fields are especially important in this connection:

- Quantification of soil functions: as storage media in the biogeochemical cycles of carbon, nitrogen, and sulfur, as well as the trace gas compounds of these elements which are responsible for climate forcing. Assessment of the potential influence on transformation processes exerted by changes in climate and land use.
- Degradation of soils due to the decoupling of element cycles through utilization. Impacts on the productivity and sustainable use of soils, and on the stability of recipient systems. Research activities at local, regional and national level.
- Effects of particulate and dissolved substances removed from soils (through erosion, eluviation, etc.) on the biotic components of neighboring limnetic and marine ecosystems (main focus on rivers, coral reefs and mangroves).
- Intensified use of remote sensing for Earth observation, and of computer simulation techniques for

describing changes in terrestrial ecosystems at regional and global level.

2.4
Biodiversity Research

Biodiversity, as a dimension of global change, is of such importance for the functions, stability and development of ecosystems that the Council considers it central to its recommendations. German biodiversity research still tends to focus too much on single disciplines and the purely national level. Wider conceptual frameworks and interdisciplinary links between the life sciences and the human sciences are still under development. The Council recommends that research be focused on the following areas:

- The basis for any assessment, preservation or restoration of biodiversity is a modern taxonomy that utilizes the methods and techniques of molecular biology more intensively, including advanced information technology. Research and educational facilities in this field are urgently in need of expansion if German researchers are to be involved in international biodiversity inventorying projects and biogeographical assessments of biodiversity.
- Research should also focus on the dichotomy between the conservation and utilization of terrestrial and aquatic ecosystems. In particular, the interrelationships between diversity, stability and function of ecosystems must be analyzed more intensively. Expanding research on population biology in order to improve nature conservation activities plays an important role in this connection. This calls for new approaches going beyond the all-too-narrow focus on biotope and species protection.
- High priority must be assigned to research into the impacts of environmental changes of varying quality, intensity and speed on populations, ecosystems and ecosystem functions (such as biogeochemical cycles). Such work should be based on findings from the areas specified above.
- Another important research field concerns the political efforts of the international community for the conservation and sustainable use of biodiversity. Research into the economics of biodiversity and the design of international environmental agreements is urgently needed.

2.5
Population, Migration and Urbanization Research

Population trends, migration and urbanization are key factors in the analysis and management of global environmental problems. Population growth and poverty are powerful driving forces behind an overall trend that is now affecting industrialized countries as well, primarily in the form of mounting migrational pressure. Research in Germany is still inadequate with respect to theoretical foundations, empirical case studies and simulation models. The following topics should be focused on:

- Rural-urban relations must be re-investigated and re-assessed, taking into account the transfers between urban areas and the subsistence economy of surrounding rural areas (reversal of the push-pull approach).
- Identifying potential sources of migration and migration flows is an increasingly important task for international migration research. In particular, systematic research must be conducted into the motivational factors driving migration.
- The determinants influencing the individual's decision to migrate must be identified in terms of sociocultural nexus and the private household context. Traditional flow analysis must be enhanced through migration system research.
- Malnutrition, undernourishment and famine are major causes of migration. Research on food security and water availability must therefore be intensified.
- The informal economy plays a central role in providing a minimum level of social security for the urban poor. In-depth research into the development potential of this sector is therefore essential.
- Our knowledge about the increasing number of megacities and large-scale agglomerations and how these operate within the global system is still fragmentary. There is also a lack of research on the informal growth of cities. To understand how "unplanned" megacities function, it is necessary to examine the systemic interrelation of these urban structures.
- The Second UN Conference on Human Settlements (HABITAT II) showed that the creation of adequate housing is an acute problem affecting the welfare of more than one billion people. Policy-oriented research should also be conducted in connection with international conferences (preparation and follow-up).

2.6
Economics Research

The Council sees a need for global economics research in the following three fields:
- *Research on the objectives and impacts of global environmental policy.* A key focus here should be the operationalization of the sustainable development principle. Above all, this requires the identification of the essential, i.e. non-substitutable elements of natural capital, the assessment of the costs of neglected environmental protection, the evaluation of intra- and intergenerational distribution, especially the scientific debate on "correct" discounting methods, and the specification of criteria regarding the economic and social compatibility of sustainable development.
- *Research on agencies responsible for global environmental policy.* A principal research focus should be the economic analysis of the behavior of globally relevant actors – both political and private (such as multinational corporations). One major issue concerns the development of strategic behavioral options which produce benefits for the overwhelming majority of those involved.
- *Research on the instruments of global environmental policy.* Due to the limited planning, regulatory and fiscal options at the global level, environmental regimes are usually implemented through treaties, conventions and economic incentives. Research on economic instruments should therefore concentrate on further development of the tradable quotas/permits option (including joint implementation), the law of liability and global funds. Another issue, parallel to these, is the question of sanction mechanisms to be applied when treaties or conventions are violated by one or more parties.

2.7
Research on Societal Organization

Research by political science on environmental topics has mainly focused on the national level, so it must now adopt a more global perspective. The problems experienced by newly-industrializing nations and their growing importance for global change deserve special attention. Policy concepts relating to global environmental protection must also take into account sociocultural and economic conditions and international law.

International environmental research within the discipline of political science must be widened in focus to embrace not only global climate issues but also other problems such as soil degradation, loss of biodiversity, and the scarcity and contamination of water resources. In view of the discrepancy between environmental awareness and the policies which are actually implemented, analysis must center, as a matter of priority, on the process of political will-formation and the implementation of international treaties. Political research must also dedicate attention to the prevention of environmental conflicts. The following tasks need to be accomplished:

- Investigation of the socioeconomic, political and cultural restrictions on action and the problems these generate for the implementation of international environmental treaties.
- Development of concepts on which to base problem-solving strategies for overcoming the typical obstacles encountered in global problem-solving processes (global commons, compliance issues, etc.).
- Analysis of the functional operation of international negotiation systems, with special reference to the uncertainty factor in global environmental change. Further, it is necessary to develop concepts for decision-making under uncertain premises.

The *discipline of law* must address global change by examining the legal options for adopting and enforcing effective measures. The legal issues include restricted national sovereignty, customary international law and ecological solidarity. Against this background, the Council recommends that the following legal issues be tackled:

- Clarification of the current body of extra-treaty standards and international customary law relating to global environmental problems, in order to react more flexibly to the latter.
- Defining a general obligation of ecological solidarity on the part of industrialized countries vis-à-vis the developing world.
- Clarification of the status of non-governmental organizations in international law.
- Clarification of legal issues concerning damages caused by global environmental change.
- Further development of enforcement mechanisms, decision-making procedures and dispute-settlement procedures in connection with international treaties.

2.8
Research on the Psychosocial Sphere

The scientific disciplines covering the psychosocial sphere are devoting increasing attention to important issues in the analysis of the causes and effects of global change and interventions to remedy the

problems which exist. This research is still little developed in Germany, with most projects involving only one discipline and decentralized organization. The following topics should be focused on, preferably by joint projects:

- Development of global change concepts for the social sciences.
- Research into guiding principles of sustainable development, from basic ethical principles to operationalization and empirical analyses.
- Studies on the determinants of environmentally relevant behavior (perception and assessment of global change phenomena, motivation, etc.) and on strategies for modifying behavior.
- Investigation and evaluation of interventions (specific contexts and groups of actors), in terms of the interactions between technical, economic, legal and psychosocial measures.
- Development, systematic application and evaluation of global change education for all levels.
- Development and establishment of a worldwide, comprehensive system of *social monitoring* (analogous to *environmental monitoring*).

Accomplishing these tasks requires more cultural and cross-cultural comparative research on social actors in the form of comprehensive, transdisciplinary case studies, whereby studies should be conducted across a wide range of spatial and temporal scales.

2.9
Technological Research

Technological research provides a key for coping with global change. A prime example is the further development of energy technologies aimed at an environmentally, economically and socially acceptable energy mix. The main focus should be placed on researching and developing different energy options, including:

- Research on *solar photovoltaics*.
- Research on the utilization of *wind energy*, especially in developing countries.

In addition, the Council recommends the promotion of research programs on the *impact of air transport on climate*, and the development of air transport along environmentally acceptable lines. At the interface between technology and economics, the Council proposes the following research topics:

- Examination of the appropriateness and effectiveness of jointly implemented activities to reduce greenhouse gases.
- Development of cost-efficient reduction strategies for greenhouse gas emissions, taking all radiatively active trace gases into account.

- Researching of techniques for removing and storing CO_2, with special reference to ecological and economic aspects.
- Analysis and quantification of the impacts of greenhouse gas emissions on the emissions of other mass contaminants of the atmosphere and other environmental problems.
- Development of cost-efficient strategies for reducing tropospheric ozone.
- Development of logistics-oriented production processes (e.g. reduction of transport pathways within the production process).
- Identification of environmentally sound industrialization processes in developing and newly-industrializing countries, taking into consideration the local technology and human resource potentials.

Practical, technology-based solutions to complex environmental problems require cooperation between various disciplines, depending on the respective project and the specific problem it addresses. The following fields should play a role in this context:

- Technologies: engineering science.
- Materials: chemistry, biology and geology.
- Planning and design: economics.
- Applications and effects: social and behavioral sciences, environmental medicine.

The recommendations formulated in the preceding section do not relate to global change research in the real sense of the term, but to research pertaining to global change within the traditional disciplines. Even disciplines focusing on planetary issues and processes, such as climate research, deliver what are in effect highly specialized bodies of knowledge that then need to be inputted into an integrated approach. In contrast, global change research should involve a constant focus on human beings as causal agents and as those affected by such change, and seeks to identify, wherever possible, the ways and means for remedying the global crisis of ecosphere-anthroposphere relations.

In order to accomplish the tasks of genuine global change research, the Council recommends that the key problems of global change – i.e. the syndromes described in Section C 2.2 – be defined as high-priority research topics. This proposal implies a realignment of environmental research strategy commensurate to the existing problems: above all, knowledge and methodologies within the various disciplines should be initiated, deployed and developed further when they can contribute to the analysis and potential management (prevention or improvement) of the syndromes as they currently operate. In *Section C 6*, we demonstrated this for the example of the *Sahel Syndrome*. This case-study must be seen as a precursor to a syndrome-based research agenda, as can be drawn up for other complexes as well.

It is neither reasonable nor feasible for German global change research to cope with all syndromes simultaneously. What is needed is a well-coordinated division of labor at international level, akin to that already existing between the various disciplines. Efforts should be made to establish a worldwide global change research program based on the concept and the logic of syndromes.

Yet there are several syndromes which the German research community should and could already be studying. The Council believes that the relevance criteria listed in *Section C 3* provide a good basis for the selection of specific syndromes. A survey conducted within the WBGU on the basis of these premises (*Section C 5*) produced an initial ranking of the syndromes. Seven complexes (listed alphabetically) were given uppermost priority:
- *Contaminated Land Syndrome.*
- *Dust Bowl Syndrome.*
- *Mass Tourism Syndrome.*
- *Sahel Syndrome.*
- *Smokestack Syndrome.*
- *Urban Sprawl Syndrome.*
- *Waste Dumping Syndrome.*

A broad-ranging discourse on the syndrome concept within the (inter)national community of global change researchers and global change decision-makers could, *inter alia*, consolidate this provisional ranking and investigate the possibility of clustering them.

In view of the interdependence of syndromes, such prioritization implies also that additional syndromes be taken into account.

The Council appeals for this discourse to be commenced as rapidly as possible, and for experiments in the interdisciplinary organization of environmental research on the basis of selected "pilot syndromes" to be conducted at once. Our specific recommendation is:
1. *to discuss and improve the syndrome concept in a series of symposia involving scientists and decision-makers from different sectors of society.* The current list of syndromes may undergo further modifications in the process;
2. *to produce an improved ranking of the syndromes based on a Delphi study;*
3. *immediately establish three research networks among existing establishments for pilot studies of*

the Smokestack, Sahel and Urban Sprawl Syndromes. These integrated studies could function as key projects under the Federal Government's new Environment Research Program.

Sections C 8 and D 4 contain further details regarding the actual organization of the discourse needed, and of the recommended networks in this connection.

The Council is well aware that a cross-sectional, problem-oriented organization of global change research could also assume different structures. However, the syndrome approach makes sense as an organizational principle given the overriding importance of sustainable development.

German research must undergo major structural improvements if it is to meet the needs of modern global change research. These include improvements to existing institutions, incentives for innovative projects, especially in tertiary education, and enhanced coordination of research and research funding. Demands for greater investment in research are frustrated by the scarcity of public funds. Lack of finance is a major obstacle, blocking further growth in research personnel and material budgets and, through non-selective staff cuts, deprives research institutions of opportunities to explore new research pathways. Shortage of public funds has imposed a restrictive framework that must be taken into consideration whenever organizational recommendations are made. The research community is therefore compelled to think about structural changes which might generate improved efficiency. Nevertheless, the organization of research in Germany has many advantages.

The strengths of a federal and pluralist structure, and the number and variety of research institutions this entails, stem from the fact that individual groups can tackle new issues flexibly and choose their own partners, especially when scientific encouragement or financial incentives are provided. On the other hand, this structure is highly intricate, which in turn hinders the concentration of research capacities under one central topic and the execution of long-term projects within international programs.

The *Wissenschaftsrat* (German Science Council) (1994) has drawn attention to the problems facing national environmental research, and has produced a set of recommendations for transdisciplinary environmental research at German universities, polytechnics and other research establishments. The obstacles are even greater for global change research, however, on account of the international context and the need to carry out investigations with foreign partners. This also explains why, in certain areas of global change research, German involvement in international programs and cooperation with developing countries is relatively confined.

Against this background, the Council puts forward a number of general organizational recommendations, grouped under three headings:
- Strengthen existing facilities and utilize approved instruments.
- Create new facilities.
- Coordinate the promotion of research.

4.1
Strengthen Existing Facilities and Utilize Approved Instruments

Existing research establishments must be given the capacity to continue ongoing projects in the field of global change research and/or to relate projects to global problems, and to start new projects involving cooperation at national and/or international level. This recommendation is directed at universities and polytechnics and to extra-university research establishments such as the Max Planck Society (MPG), the Helmholtz research association (HGF), the "Blue List" research institutes (WBL) and the Fraunhofer Society (FhG), as well as the research facilities operated by certain federal agencies. Impulses in this direction must come from the facilities themselves or from the bodies which operate and control them, i.e. by redefining the priorities and content of research and by organizational restructuring.

What is absolutely essential, however, is the use of tried and tested support instruments on the part of the BMBF (joint projects, *research networks*) and the DFG (*priority programs*, *collaborative research centers*). *Research groups* and *graduate colleges* are suitable instruments, whereby the restrictive principle that research units must be sited at a single location should be relaxed in view of the technical opportunities provided by modern telecommunications.

All these integrating measures should also be applied in the *education and training* of domestic and foreign students and prospective scientists. Aspects of global change should be referred to during basic level courses, and studied in greater detail in more advanced courses.

Major items of research equipment are absolutely essential in many areas of global change research. These include equipment for remote sensing and climate research using supercomputers, research vessels, remote sensing satellites and monitoring stations. Global change research also needs large-scale, comprehensive and long-term *observation data* on the environment, the economy and sociocultural aspects. It relies on comparisons between *cultures* and *ecosystems* and must build on detailed and broadly conceived *case studies* as well as complex *models*. The Council attaches considerable importance to ensuring continued provision of these basic requirements.

Germany's participation in international programs varies in quality, and in some important areas is in need of expansion. We recommend continued involvement in *international institutions* and *secretariats*, in terms of input, staffing and financial contribution, whereby greater integration of German researchers by such institutions would be desirable.

4.2
Create New Facilities

The Council recommends the establishment of a *German Strategy Center on Global Change* in order to enhance problem-solving capacity with respect to global change and to strengthen interdisciplinary cooperation. The Center would carry out complex analyses, using external experts as well to provide scientific support for decision-making processes. It should take up the suggestions of policymakers in the political and public spheres and translate these into research issues, as well as process existing knowledge to aid decision-making processes in politics, industry and society in general.

The Council believes that *small research centers* should be set up at or around universities for limited periods of time; these would work on acute problems in the field of global change research over 10 years or so, and ensure German participation in international programs.

In addition, the Council recommends the creation of *research networks* as long-term, purpose-made alliances between independent research institutions for joint work on complex issues, such as a specific syndrome, and for further refinement of methodologies. These should include the use of modern technologies for data acquisition, storage and transmission within national and international frameworks.

Responsible research bodies (MPG, HGF, WBL and FhG), the DFG, the BMBF and specialized research establishments and university departments should jointly create flexible institutions to deal with specific global change problems (*inter-institutional research*).

The Council appeals to industry and commerce, especially the multinational corporations, to set up a *"Global Change" Foundation* as an expression of environmental self-commitment. Such a body would help compensate for the financial restrictions referred to earlier. The Foundation should promote a dialog between the scientific community, economic policymakers and the media on global change issues. It could also be present at the EXPO 2000 World Exhibition in Hanover, Germany.

4.3
Coordinate the Promotion of Research

The two most important institutions providing funding and support for research in Germany are the BMBF and the DFG. The BMBF has several ministerial departments and various project support units responsible for specific research fields relating to global change. The same is true of the DFG, which is organized according to scientific disciplines. Both institutions must strengthen their efforts towards transdisciplinary planning and assessment. There is also a need for closer coordination between the DFG and the BMBF regarding the deployment of instruments for promoting global change research.

Within the Federal Government, supervisory control of global change research is not confined to the BMBF. Although the BMU does not operate its own research establishments, it supports a number of global change research projects through the Federal Environment Agency (UBA). Major research facilities and projects are also operated by the Federal Ministry of Transport (BMV), the Federal Ministry of the Economy (BMWi), the Federal Ministry for Food, Agriculture and Forestry (BML), the Federal Ministry of Economic Cooperation and Development (BMZ) and the Federal Ministry of the Interior (BMI). The Council sees a need for coordination here, which should go beyond the work of the Interministerial Research Group (IMA) on Global Change.

The Council is monitoring with great interest the efforts of the DFG to establish a German Global Change Committee, comprising functional units of the Senate Committee for Environmental Research (SAUF) and the German IGBP Committee, for the purpose of planning and supporting research involvement in international global change programs. This National Committee could also play a role in coordinating the various global change research activities in Germany.

The Council also recommends that the Federal Chancellery produce an integrated *Global Report* in the middle of each legislature period. In view of the developments triggered off by UNCED in Rio de Janeiro, the Report should provide information on Federal Government activities concerning global change and sustainable development. Policymaking and research activities in Germany should be analyzed in terms of their environmental, economic and sociocultural impacts within the global network of interrelations. The Council firmly believes that such a report would become an important source of information for the general public in Germany and for foreign institutions, and that it would exert a consolidating and integrating influence on global change activities in the various federal ministries.

The work of the German Parliament's Enquete Commissions has had an integrating effect on German research and government support agencies. An *Enquete Commission on Global Change* could continue on the work of the current Enquete Commission Protection of People and the Environment, whereby the focus should be placed on implementing the recommendations of the scientific community as advanced, for example, by the German Advisory Council on Global Change.

For some time now, there have been discussions about establishing a German Academy of the Sciences, similar to those in other countries, which could state its position on issues of national importance with a high degree of independence and authority; were such an academy to be created, the problem of global change would certainly be an important topic for it to consider.

5   Prospects

In the Introduction to this Report, the Council laid stress on the fact that research cannot be seen as a substitute for political action, but instead is a prerequisite for appropriate measures to protect the Earth System and for sustainable use of its resources. Postponing environmental measures until there is scientific proof of their necessity is just as weak a response as political actionism devoid of scientific foundations.

The recommendations made in this year's Annual Report regarding the further development of the German research effort are aimed at ensuring the necessary scientific basis for decision-making in the field of environmental policy and at developing methods with which measures taken can be provided with critical support, controlled and assessed in terms of their primary and secondary impacts. The scope and complexity of global change phenomena render it almost impossible to conduct scientific experiments in this field. Instead, research relates primarily to the analysis of long series of observed data, comparative case studies, the synthesis of existing data records and knowledge, drawing analogies from small-scale processes to large-scale, global phenomena, and above all computer simulations using complex models constructed using knowledge and theoretical analysis of processes and interactions within the Earth System.

The scientific fascination of global change research lies in the inspired synthesis of the most diversified knowledge about processes and interactions in the ecosphere and anthroposphere. This gives rise to encounters on a new, higher plane between the natural sciences, the social sciences and the humanities, which for the last two centuries have become increasingly alienated from each other.

The social fascination of global change research lies in the fact that it creates a crucial basis for serving the long-term needs of humanity in the complex Earth System, i.e. creating the basis for intergenerational equity.

Some topics of special relevance for global change research have not been considered in this Annual Report. They include, for example, the field of health and environment, which deserves greater attention in the analysis of the impacts of global change on society than is dedicated to it here. More attention should also be devoted to exploring the role of the media with regard to the perception and assessment of global change problems. The Report does not deal, for example, with the importance of theological or historical analyses of people-environment relations in different cultures and epochs of human history.

In relation to its population, Germany bears a disproportionate responsibility for the causes of global change. Its contribution towards global change research, albeit substantial, must be radically increased. The primary requirement is not so much a major increase in research funding, or the founding of new large-scale research establishments, but the efficient use of data and knowledge already available and the synthesis of that knowledge to solve complex problems (e.g. syndrome research). What is also needed are organizational measures to ensure that existing global change research potential is deployed more effectively, and that gaps in the various research fields can be closed by providing a modest level of extra funding.

Transnational networking and integration into international programs at European and global level are crucially important for German global change research. Commensurate with Germany's role in the world economy, German research should play a leading role in creating and expanding research capacities in the developing countries.

References E

Andresen, S., Skjaseth, J. B. and Wettestad, J. (1995): Regime, the State and Society – Analysing the Implementation of International Environmental Commitments. Laxenburg: IIASA.

Barrett, S. (1991): The Paradox of International Environmental Agreements. Mimeo, London: London Business School and Centre for Social and Economic Research on the Global Environment.

Barrett, S. (1993): Joint Implementation for Achieving National Abatement Commitments in the Framework Convention on Climate Change. Revised Draft for Environment Directorate Organisation for Economic Cooperation and Development. London: London Business School and Centre for Social and Economic Research on the Global Environment.

Bateman, I. J. and Turner, R. K. (1993): Valuation of the Environment, Methods and Techniques: The Contingent Valuation Method. In: Turner, R. K. (ed.): Sustainable Environmental Economics and Management: Principles and Practice. London, New York: Belhaven, 120-191.

Beirat für Naturschutz und Landschaftspflege beim BMU (1995a): Naturschutzforschung und -lehre: Situation und Forderungen. Natur und Landschaft 70 (1), 5-10.

Beirat für Naturschutz und Landschaftspflege beim BMU (1995b): Zur Akzeptanz und Durchsetzbarkeit des Naturschutzes. Natur und Landschaft 70 (2), 51-61.

Bisby, F. A. (1995): Characterization of Biodiversity. In: Heywood, V. H. and Watson, R. T. (eds.): Global Biodiversity Assessment. Cambridge: Cambridge University Press, 21-106.

BMFT – Bundesministerium für Forschung und Technologie (1990): Bericht des Forschungsbeirats Waldschäden/Luftverunreinigungen über den Zeitraum 1987/89. Kurzfassung. Bonn: BMFT.

BMU – Bundesministerium für Umwelt Naturschutz und Reaktorsicherheit (1992): Konferenz der Vereinten Nationen für Umwelt und Entwicklung im Juni 1992 in Rio de Janeiro – Dokumente. Bonn: BMU.

BMZ – Bundesministerium für wirtschaftliche Zusammenarbeit (1995): Das Sektorkonzept. Umweltgerechte Kommunal- und Stadtentwicklung. BMZ aktuell 058, 8-12.

Brenck, A. (1992): Moderne umweltpolitische Konzepte: Sustainable Development und ökologisch-soziale Marktwirtschaft. Zeitschrift für Umweltpolitik & Umweltrecht 14 (4), 379-413.

Brown-Weiss, E. (1992): Environmental Change and International Law. New Challenges and Dimensions. Tokyo: United Nations University Press.

BUND – Bund für Umwelt und Naturschutz und Misereor (1996): Zukunftsfähiges Deutschland. Ein Beitrag zu einer global nachhaltigen Entwicklung. Basel: Birkhäuser.

CGCP – Canadian Global Change Program (1993): Canadian Involvement in International Global Change Activities. A Compendium (CIGA). Ottawa, Ontario: CGCP and Royal Society of Canada.

Conrad, K. and J. Wang (1993): Quantitative Umweltpolitik: Gesamtwirtschaftliche Auswirkungen einer CO_2-Besteuerung in Deutschland (West). Jahrbücher für Nationalökonomie und Statistik 212, 309-324.

Dasgupta, P. (1995): Bevölkerungswachstum, Armut und Umwelt. Spektrum der Wissenschaft (7), 54-59.

Dechema – Deutsche Gesellschaft für Chemisches Apparatewesen (1994): Kurzfassungen der Vortragsgruppen auf der Ausstellung chemischer Apparaturen (ACHEMA). Frankfurt: Dechema.

DFG – Deutsche Forschungsgemeinschaft (1992): Perspektiven der Forschung und ihrer Förderung. Aufgaben und Finanzierung 1993-1996. Bonn: DFG.

Diversitas (1995): DIVERSITAS. An International Programme of Biodiversity Science. The Next Phase. Paris: IUBS, SCOPE, UNESCO, ICSU, IGBP-GCTE, IUMS.

DIW – Deutsches Institut für Wirtschaftsforschung (1994): Ökosteuer – Sackgasse oder Königsweg? Studie im Auftrag von Greenpeace. Berlin: DIW.

EA – Environment Agency of Japan (1995): Global Environment Research in Japan. Tokyo: EA.

Ebenhöh, W., Sterr, H. and Scimmering, F. (1995): Küsten im Klimawandel. Einblick 22, 9-16.

Endres, A. and Schwarze, R. (1994): Das Zertifikatemodell in der Bewährungsprobe? Eine ökonomische Analyse des "Acid Rain"-Programms des neuen US-Clean Air Acts. In: Endres, A., Rehbinder, E. and Schwarze, R. (eds.): Umweltzertifikate und Kompensationen in ökonomischer und juristischer Sicht. Bonn: Economica, 137-215.

Enquete Commission "Protection of People and Environment" of the 12th German Bundestag (1994): Responsibility for the Future. Options for Sustainable Management of Substance Chains and Material Flows. Bonn: Economica.

Enquete Commission "Protecting the Earth's Atmosphere" of the 12th German Bundestag (1995a): Securing our Earth's Future. Long-term Global Warming Management through Sustainable Energy Policies. Bonn: Economica.

Enquete Commission "Protection of People and Environment" of the 12th German Bundestag (1995b): Shaping Industrial Society. Prospects for Sustainable Management of Substance Chains and Material Flows. Bonn: Economica.

Erdmann, K.-H. and Nauber, J. (1995): Der deutsche Beitrag zum UNESCO-Programm "Der Mensch und die Biosphäre" (MAB) im Zeitraum Juni 1992 bis Juli 1994. Bonn: BMU.

European Commission (1994): Das 4. Rahmenprogramm. Allgemeine Information. Brussels: European Commission.

EWI – Energiewirtschaftliches Institut Köln (1994): Gesamtwirtschaftliche Auswirkungen von Emissionsminderungsstrategien. Studie im Auftrag der Enquete-Kommission "Schutz der Erdatmosphäre". Cologne

Freeman III, A. M. (1993): The Measurement of Environmental and Resource Values. Washington, DC: Resources for the Future.

Fromm, O. and Hansjürgens, B. (1994): Umweltpolitik mit handelbaren Emissionszertifikaten – eine ökonomische Analyse des RECLAIM-Programms in Südkalifornien. Zeitschrift für angewandte Umweltforschung 7 (2), 211-223.

Gerken, L. (1995): Competition Among Institutions. London: McMillan.

Gerken, L. and Renner, A.(1995 bzw. 1996): Ordnungspolitische Grundfragen einer Politik der Nachhaltigkeit. Studie im Auftrag des Bundeswirtschaftsministeriums. Freiburg.

Görres, A., Ehringhaus, H. and von Weizsäcker, E. U. (1994): Der Weg zur ökologischen Steuerreform. Memorandum des Fördervereins ökologische Steuerreform. Munich: Olzog.

Gray, P. C. R. (1995): Social Science Research in the United Kingdom into Global Environmental Change. Jülich: Programmgruppe MUT.

GTZ – Gesellschaft für Technische Zusammenarbeit GmbH (1995): Tropenökologisches Begleitprogramm (TÖB). Ziele, Konzeption und Vergabekriterien. Eschborn: GTZ.

Hansjürgens, B. and Fromm, O. (1994): Erfolgsbedingungen von Zertifikatelösungen in der Umweltpolitik – am Beispiel der Novelle des US-Clean Air Act von 1990. Zeitschrift für Umweltpolitik & Umweltrecht 17 (4), 473-505.

Hauser, J. A. (1990): Bevölkerungs- und Umweltprobleme in der Dritten Welt. Band 1. Bern, Stuttgart: Paul Haupt.

Hauser, J. A. (1991): Bevölkerungs- und Umweltprobleme in der Dritten Welt. Band 2. Bern, Stuttgart: Paul Haupt.

Heister, J., Michaelis, P. and Mohr, E. (1991): Umweltpolitik mit handelbaren Emissionsrechten: Möglichkeiten zur Verringerung der Kohlendioxid- und Stickoxidemissionen. Tübingen: J.C.B. Mohr.

Heister, J. and Stähler, F. (1995): Globale Umweltpolitik und Joint Implementation: Eine ökonomische Analyse für die Volksrepublik China. Zeitschrift für Umweltpolitik & Umweltrecht 18 (2), 205-230.

Henle, K. and Kaule, G. (1992): Arten- und Biotopschutzforschung für Deutschland. In: Forschungszentrum Jülich (ed.): Berichte aus der Ökologischen Forschung. Linnich: WEKA- Druck, 435.

Heywood, V. (1993): Die neue allumfassende Wissenschaft. Naturopa 73, 4-5.

Heywood, V. H. and Baste, I. (1995): Introduction. In: Heywood, V. H. and Watson, R. T. (eds.): Global Biodiversity Assessment. Cambridge: Cambridge University Press, 1-19.

Heywood, V. H. and Watson, R. T. (1995): Global Biodiversity Assessment. Cambridge: Cambridge University Press.

Huckestein, B. (1993): Umweltlizenzen – Anwendungsbedingungen einer effizienten Umweltpolitik durch Mengensteuerung. Zeitschrift für Umweltpolitik & Umweltrecht 16 (1), 1-29.

IACGEC – Inter-Agency Committee on Global Environmental Change (1993): The UK Research Framework 1993. Swindon: IACGEC.

IACGEC – Inter-Agency Committee on Global Environmental Change (1996): UK National Strategy for Global Environmental Research 1996. Draft for Discussion at IACGEC National Strategy Consultation Meeting. Swindon: IACGEC.

IAW – Institut für angewandte Wirtschaftsforschung (1995): Ordnungspolitische Aspekte des Nachhaltigkeitsanliegens. Unveröffentlichtes Gutachten. Tübingen.

IGBP – International Geosphere Biosphere Programme (1994): IGBP in Action: The Work Plan 1994-1998. Stockholm: IGBP.

IÖW – Institut für ökologische Wirtschaftsforschung (1995): Ordnungspolitische Aspekte des Nachhaltigkeitsanliegens. Unveröffentlichtes Gutachten. Berlin.

IPCC – International Panel on Climate Change (1990): Climate Change. The IPCC Scientific Assessment. Cambridge, New York, Melbourne: Cambridge University Press.

IPCC – International Panel on Climate Change (1992): Climate Change 1992. The Supplementary Report to the IPCC Scientific Assessment. Cambridge, New York, Melbourne: Cambridge University Press.

IPCC – International Panel on Climate Change (1996): Climate Change 1995. The Second Assessment Report of the IPCC. Cambridge, New York, Melbourne: Cambridge University Press.

Joußen, W. (1995): Human Dimensions in Global Environmental Research in den Niederlanden. Organisations- und Themenstruktur – Kurzbericht. Eschweiler: Büro für sozialwissenschaftliche Analysen und Planungen.

Karadeloglu, P. (1992): Energy Tax versus Carbon Tax: A Quantitative Macroeconomic Analysis with the Hermes/Midas Models. In: Commission of the European Communities (ed.): European Economy, The Economics of Limiting CO_2 Emissions (special edition 1), 153-184.

Karger, C. R. (1992): Global Environmental Change: Deutsche und internationale Forschungsprogramme zu "Human Dimensions of Global Environmental

Change". Jülich: Programmgruppe MUT.

Kaule, G. and Henle, K. (1992): Forschungsdefizite im Aufgabenbereich des Arten- und Biotopschutzes. Jahrbuch für Naturschutz und Landschaftspflege 45, 127-136.

Kleinkauf, W., Sachhau, J. and Hempel, H. (1993): Modulare Energieaufbereitung und Anlagentechnik. Strategische Ansätze zur Gestaltung PV-gerechter Systemtechnik. In: Forschungsverbund Sonnenenergie (ed.): Themen 92/93. Cologne: Verlag Photovoltaik, 9-16.

Klemmer, P. (1994): Ressourcen- und Umweltschutz um jeden Preis. In: Voss, G. (ed.): Sustainable Development – Leitziel auf dem Weg in das 21. Jahrhundert. Kölner Texte und Thesen 17, 22-57.

Klemmer, P. (im Druck): Das Prinzip der Nachhaltigkeit: Neuere stoffpolitische Ansätze. Bochum: List Gesellschaft.

Kreibich, V. (1992): Stadtentwicklung in Afrika – die Auflösung der Stadt? In: Deutsche Gesellschaft für die Vereinten Nationen (ed.): Megastädte. Zeitbomben mit globalen Folgen. Blaue Reihe (44), 22-31.

Krumm, R. (1996): Internationale Umweltpolitik. Eine Analyse aus umweltökonomischer Sicht. Heidelberg, Berlin: Springer.

Lind, R. C. (1990): Reassessing the Government's Discount Rate Policy in Light of New Theory and Data in a World Economy with a High Degree of Capital Mobility. Journal of Environmental Economics and Management 18, 8-28.

Maier-Rigaud, G. (1994): Umweltpolitik mit Mengen und Märkten: Lizenzen als konstituierendes Element einer ökologischen Marktwirtschaft. Marburg: Metropolis.

Markl, H. (1995): Wohin geht die Biologie? Biologie in unserer Zeit 25 (3), 33-39.

McLeod, D. (1995): Global Change Research Themes. A Report of the Canadian Global Change Program Research Committee. Ottawa, Ontario: The Royal Society of Canada.

Mertins, G. (1992): Urbanisierung, Metropolisierung und Megastädte. Ursachen der Stadt"explosion" in der Dritten Welt – Sozioökonomische und ökologische Problematik. In: Deutsche Gesellschaft für die Vereinten Nationen (ed.): Megastädte. Zeitbomben mit globalen Folgen. Blaue Reihe (44), 7-21.

Michaelowa, A. (1995): Internationale Kompensationsmöglichkeiten zur CO_2-Reduktion unter Berücksichtigung steuerlicher Anreize und ordnungsrechtlicher Maßnahmen. Hamburg: HWWA-Institut für Wirtschaftsforschung,

Mitchell, R. C. and Carson, R. T. (1989): Using Surveys to Value Public Goods: The Contingent Valuation Method. Washington, DC: Resources for the Future.

Mittelstraß, J. (1989): Wohin geht die Wissenschaft? Über Disziplinarität, Transdisziplinarität und das Wissen in einer Leibniz-Welt. Konstanzer Blätter für Hochschul-

fragen 26 (1-2), 97-115.

MuD – Meß- und Dokumentationsprogramm des BMBF, WIP – Wirtschafts- und Infrastruktur-Planungs KG und Lehrstuhl für angewandte Physik der Universität Cottbus (1996): Statusreport Photovoltaik. Umweltmagazin (April), 76-78.

Nelson, R. (1988): Dryland Management: The "Desertification" Problem. Washington, DC: The World Bank.

NOAA – National Oceanic and Atmospheric Administration (1993): Report of the NOAA Panel on Contingent Valuation. Federal Register 58 (10), 4602-4614.

NRP Programme Office (1994): Dutch National Research Programme on Global Air Pollution and Climate Change. Main Features Second Phase 1995-2000. Bilthoven: NRP.

OECD – Organization for Economic Co-operation and Development (1994): Project and Policy Appraisal: Integrating Economics and Environment. Paris: OECD.

Oliveira-Martins, J. (1992); The Costs of Reducing CO_2-Emissions: A Comparison of Carbon Tax Curves with GREEN. OECD-Working-Papers. Paris: OECD.

Panzer, G. (1995): Anstöße geben. Politische Ökologie 13 (Sonderheft 7), 10-14.

Pruckner, G. J. (1995): Der kontingente Bewertungssatz zur Messung von Umweltgütern. Stand der Debatte und umweltpolitische Einsatzmöglichkeiten. Zeitschrift für Umweltpolitik & Umweltrecht 18 (4), 503-536.

Rawls, J. (1972): A Theory of Justice. Oxford: Clarendon Press.

Rentz, H. (1995a): "Joint Implementation" in der internationalen Umweltpolitik. Zeitschrift für Umweltpolitik & Umweltrecht, 18 (2), 179-204.

Rentz, H. (1995b): Kompensationen im Klimaschutz: ein erster Schritt zu einem nachhaltigen Schutz der Erdatmosphäre, Berlin: Duncker + Humblot.

RIVM – National Institute of Public Health and Environmental Protection (1993): RIVM Global Change Research Programme. An Overview. Bilthoven: RIVM.

RMNO – Advisory Council for Research on Nature and Environment (1996): Research Activities on Nature and Environment. Overview of National and International Programmes and Organizations. Rijswijk: RMNO.

RWI – Rheinisch-Westfälisches Institut für Wirtschaftsforschung (1995): Gesamtwirtschaftliche Beurteilung von CO_2-Minderungsstrategien. Gutachten im Auftrag des BMWi. 2. Zwischenbericht. Essen: RWI.

SGCR – Subcommittee on Global Change Research und NSTC – Committee on Environment and Natural Resources Research of the National Science and Technology Council (1996): Our Changing Planet. The Fiscal Year 1996 U.S. Global Change Research Programme. An Investment in Science for the Nation's Future. Washington, DC: SGCR und NSTC.

Solbrig, O. T. (1991): Biodiversität: Wissenschaftliche Fragen und Vorschläge für die internationale Forschung. Bonn: Rheinischer Landwirtschafts-Verlag.

SPP – Schwerpunktprogramm Umwelt (1994): Übersicht. Bern: SPP.

SPP – Schwerpunktprogramm Umwelt (1995): Ausführungsplan zum Schwerpunktprogramm Umwelt (SPP). Beitragsperiode 1996-1999. Bern: SPP.

SRU – Rat von Sachverständigen für Umweltfragen (1994): Umweltgutachten 1994. Für eine dauerhaft-umweltgerechte Entwicklung. Stuttgart: Metzler-Poeschel.

SRU – Rat von Sachverständigen für Umweltfragen (1996): Umweltgutachten 1996. Zur Umsetzung einer dauerhaft-umweltgerechten Entwicklung. Stuttgart: Metzler-Poeschel.

Standaert, S. (1992): The Macro-Sectoral Effects of an EC-wide Energy Tax: Simulation Experiments for 1993-2005. In: Commission of the European Communities (ed.): European Economy. The Economics of Limiting CO_2 Emissions (special edition 1), 84-98.

Stork, N. E. and Samways, M. J. (1995): Inventoring and Monitoring of Biodiversity. In: Heywood, V. H. and Watson, R. T. (eds.): Global Biodiversity Assessment. Cambridge: Cambridge University Press, 453-543.

Sukopp, H. (1992): Training and Research: University Basic Curricula. In: Swiss National Commission for UNESCO (ed.): Education and Science for Maintaining Biodiversity. Basel, Paris: UNESCO, 66-74.

Thomas, D. S. G. and Middleton, N. J. (1994): Desertification. Exploding the Myth. Chichester, New York: John Wiley & Sons.

UBA – Umweltbundesamt (1992): Umweltforschungskatalog – UFOKAT '92. Berlin: E. Schmidt.

UBA – Umweltbundesamt (1994): Jahresbericht des Umweltbundesamtes. Berlin: UBA.

UGR – Beirat Umweltökonomische Gesamtrechnung (1995): Zweite Stellungnahme des Beirats "Umweltökonomische Gesamtrechnung" beim Bundesminister für Umwelt, Naturschutz und Reaktorsicherheit zu den Umsetzungskonzepten des Statistischen Bundesamtes. Zeitschrift für angewandte Umweltforschung 8 (4), 455-476.

UK Global Environmental Research (GER) Office (1996): Directory of Global Environmental Research Programmes and Contact Points in the UK. Swindon: UK GER Office.

UN – United Nations (1994): International Conference on Population and Development (ICPD). Konferenzdokumente. Geneva: UN.

UN – United Nations (1995): World Urbanization Prospects, the 1994 Revision. New York: UN

UN – United Nations (1996): United Nations Conference on Human Settlements (Habitat II). The Habitat Agenda: Goals and Principles, Commitments and Global Plan of Action. Geneva: UN.

UNDP – United Nations Development Programme (1992): Human Development Report 1992. Oxford: Oxford University Press.

Unwin, T. and Potter, R. B. (1989): Urban-Rural Interaction in Developing Countries – A Theoretical Perspective. In: Potter, B. and Unwin, T. (eds.): The Geography of Urban-Rural Interaction in Developing Countries. London, New York: Routledge, 11-32.

WBGU – German Advisory Council on Global Change (1993): World in Transition: Basic Structure of Global Human-Environment Interactions. 1993 Annual Report. Bonn: Economica.

WBGU – German Advisory Council on Global Change (1995a): World in Transition: The Threat to Soils. 1994 Annual Report. Bonn: Economica.

WBGU – German Advisory Council on Global Change (1995b): Scenario for the derivation of global CO_2 reduction targets and implementation strategies. Statement on the occasion of the First Conference of the Parties to the Framework Convention on Climate Change in Berlin. Bremerhaven: WBGU.

WBGU – German Advisory Council on Global Change (1996): World in Transition: Ways Towards Global Environmental Solutions. 1995 Annual Report. Berlin, Heidelberg, New York: Springer.

WCMC – World Climate Monitoring Center (1992): Global Biodiversity: Status of the Earth's Living Resources. London, Glasgow, New York: Chapman & Hall.

Welsch, H. and F. Hoster (1995): A General Equilibrium Analysis of European Carbon/Energy Taxation: Model Structure and Macroeconomic Results. Zeitschrift für Wirtschafts- und Sozialwissenschaften 115 (2), 211-235

Wissenschaftsrat (1994): Stellungnahme zur Umweltforschung in Deutschland. Köln: Wissenschaftsrat.

WRI – World Resources Institute (1996): World Resources 1996-97. The Urban Environment. New York, Oxford: Oxford University Press.

ZEW – Zentrum für europäische Wirtschaftsforschung (1995): Ordnungspolitische Aspekte des Nachhaltigkeitsanliegens. Unveröffentlichtes Gutachten. Mannheim.

Ziegler, W., Bode, H.-J., Mollenhauer, D., Peters, D. S., Schminke, H. K., Trepl, L., Türkay, M., Zizka, G. and Zwölfer, H. (1996): Biodiversität. Entwurf einer Denkschrift für die Deutsche Forschungsgemeinschaft. Bonn: Arbeitsgruppe des Senatsausschusses für Umweltforschung.

Zimmermann, H. (1992): Ökonomische Aspekte globaler Umweltprobleme. Zeitschrift für angewandte Umweltforschung 5, 310-322.

Glossary F

Collaborative research centers

Collaborative research centers are special research establishments set up with DFG support for longer-term periods (12-15 years and more), where researchers from a variety of disciplines collaborate within the framework of a multidisciplinary research program. They are each located at a specific university, whereby several neighboring universities and extra-university facilities may be involved, as may industrial and commercial enterprises.

Core problems of global change

The core problems are those global change phenomena currently of central importance. In the ⇨syndrome approach they appear either as particularly dominant ⇨trends of global change (e.g. climate change, population growth), or comprise several interrelated trends. One such "megatrend" is the core problem of "soil degradation", which involves several different trends such as soil erosion, salinization, contamination, etc.

"Crash barrier" concept

The concept of "crash barrier" demarcates the domain of free action for the people-environment system from those domains which represent undesirable or even catastrophic developments and which must be avoided at all costs. Pathways for sustainable development run within the corridor defined by these crash barriers.

Disposition

Disposition refers to the vulnerability of a region to a specific ⇨syndrome. The disposition space, meaning the geographical distribution of the disposition, is determined by natural and anthropospheric conditions exhibiting characteristics that are structural in nature, i.e. which can only be altered over long time-scales.

Environmental monitoring

Environmental monitoring refers to the observation, recording and analysis of environmental states and changes in same. Comprehensive surveillance networks for climate, ocean, freshwater, land use, etc. ensure that global trends and changes are registered and disseminated to politicians and researchers. Environmental monitoring programs are currently coordinated and conducted by international organizations such as the WMO and UNEP.

Exposition

Exposition refers to natural and anthropogenic events and processes that can trigger a ⇨syndrome and are mostly of a short-term nature (e.g. sudden natural disasters, rapid changes in exchange rates, etc.). The syndrome mechanism is triggered off by them in vulnerable regions (Disposition).

Global network of interrelations

The global network of interrelations is a qualitative network embracing all ⇨trends of global change and their interactions. It provides a highly aggregated description of the global change system.

Graduate colleges

Graduate colleges are university institutions for promoting postgraduate researchers (doctoral candidates). The intention is to enable the latter to work on their doctoral theses within a systematically structured and transdisciplinary study program related to a joint research program conducted by the various professors involved in the scheme.

Interdisciplinarity

Interdisciplinarity refers to cooperation between different disciplines, at least during certain phases. In contrast to ⇨multidisciplinarity, therefore, interdisciplinarity involves more than the mere "addition" of single disciplines, because in a preliminary phase the problems to be dealt with are discussed and identified on a joint basis, and the results brought together in an integral whole on completion of activities. The individual aspects of the research topic, however, continue to be analyzed with the respective disciplinary methods.

Multidisciplinarity

Multidisciplinary research is characterized by different research disciplines working on the same research issue largely independently of each other.

Network projects

Network projects are an instrument developed by the BMBF for promoting research. They are affiliated to existing research institutions and foster cooperation between researchers and practitioners through involvement in the programs on the part of independent research institutes and local government corporations as well as universities. They are set up for a period of several years.

Priority programs

Priority programs are research promotion procedures specific to the DFG and are generally established for a term of five years. A special feature of a priority research program is the collaboration at national level of participant researchers who are free, within pre-defined thematic boundaries, in their selection of topic, research agenda and methodologies.

Research networks

Research networks are long-term "purpose-made" alliances between independent scientific establishments aimed at collaboration on complex, usually interdisciplinary issues and at further development of methodological foundations.

Social monitoring

Social monitoring is the long-term surveillance of societal trends. It encompasses continuous or periodic economic monitoring and description (e.g. national accounts) as well as social and behavioral science programs for repeated surveying of attitudes, opinions, knowledge levels and valuation among the general population, for example.

Sustainable development

A somewhat diffuse term for which there are various definitions, translations and interpretations. It refers to a concept for environment and development that was first elaborated in the Brundtland Report and further refined at the 1992 UN Conference for Environment and Development in Rio de Janeiro. The syndrome concept used by the WBGU is a means for operationalizing this weakly defined term (*see Section C 2.1.2*).

Syndromes of global change

Syndromes are functional patterns of people-environment relations, in other words characteristic negative constellations of natural and anthropogenic ⇨trends of global change and their respective interactions. Each syndrome - or "clinical profile", to use a medical analogy - represents an anthropogenic cause-effect complex involving specific environmental stresses, and an independent pattern of environmental degradation. Syndromes are transsectoral in nature, i.e. they affect several sectors (such as the economy, the biosphere, population) or environmental media (soils, water, air), yet are always related, directly or indirectly, to natural resources. Syndromes can usually be identified in different forms in many regions of the world, whereby several syndromes may occur simultaneously.

Transdisciplinarity

Transdisciplinary research goes beyond the confines of specific disciplines and defines and manages its objects of knowledge independently of such disciplines. Models and methodologies used within the individual disciplines are examined critically for their aptitude for the respective research topic, or new methods are developed.

Trends of global change

Trends are both anthropospheric and natural phenomena that are relevant for and characterize global change. They represent variable or processual factors that can be determined at least qualitatively. Examples of trends include population growth, the anthropogenic greenhouse effect, growing environmental awareness and medical advances.

The German Advisory Council on Global Change

G

Prof. Horst Zimmermann, Marburg
(Chairperson)
Prof. Hans-Joachim Schellnhuber, Potsdam
(Vice Chairperson)
Prof. Friedrich O. Beese, Göttingen
Prof. Gotthilf Hempel, Bremen
Prof. Lenelis Kruse-Graumann, Hagen
Prof. Paul Klemmer, Essen
Prof. Karin Labitzke, Berlin
Prof. Heidrun Mühle, Leipzig
Prof. Udo Ernst Simonis, Berlin
Prof. Hans-Willi Thoenes, Wuppertal
Prof. Paul Velsinger, Dortmund

STAFF TO THE COUNCIL MEMBERS
Dr. Arthur Block, Potsdam
Dipl.-Ing. Sebastian Büttner, Berlin
Dr. Svenne Eichler, Leipzig
Dipl.-Volksw. Oliver Fromm, Marburg
Dipl. Psych. Gerhard Hartmuth, Hagen
Dipl.-Met. Birgit Köbbert, Berlin
Dipl.-Geol. Udo Kubitz, Essen
Dr. Gerhard Lammel, Hamburg
Dipl.-Volksw. Wiebke Lass, Marburg
Dipl.-Ing. Roger Lienenkamp, Dortmund
Dr. Heike Schmidt, Bremen
Dr. Rüdiger Wink, Bochum
Dr. Ingo Wöhler, Göttingen

THE SECRETARIAT OF THE COUNCIL,
BREMERHAVEN*
Prof. Meinhard Schulz-Baldes
(Director)
Dr. Carsten Loose
(Deputy Director)
Heinke Deloch, M. A.
Vesna Karic
Ursula Liebert
Dr. Benno Pilardeaux
Dipl.-Volksw. Barbara Schäfer
Martina Schneider-Kremer, M.A.

*   Secretariat WBGU
    c/o Alfred-Wegener-Institute for Polar and
    Marine Research
    PO Box 12 01 61
    D-27515 Bremerhaven
    Germany

    Phone: ++49-471-4831-723
    Fax: ++49-471-4831-218
    Email: wbgu@awi-bremerhaven.de
    Internet: http://www.awi-bremerhaven.de/WBGU/

Joint Decree on the Establishment of the German Advisory Council on Global Change (April 8, 1992)

Article 1

In order to periodically assess global environmental change and its consequences and to help all institutions responsible for environmental policy as well as the public to form an opinion on these issues, an Advisory Council on "Global Environmental Change" reporting to the Federal Government shall be established.

Article 2

(1)
The Council shall submit a report to the Federal Government by the first of June each year, giving an updated description of the state of global environmental change and its consequences, specifying quality, size and range of possible changes and giving an analysis of the latest research findings. In addition, the report should contain indications on how to avoid or correct maldevelopments. The report shall be published by the Council.

(2)
While preparing the reports, the Council shall provide the Federal Government with the opportunity to state its position on central issues.

(3)
The Federal Government may ask the Council to prepare special reports and opinions on specified topics.

Article 3

(1)
The Council shall consist of up to twelve members with special knowledge and experience regarding the tasks assigned to the Council.

(2)
The members of the Council shall be jointly appointed for a period of four years by the two ministries in charge, the Federal Ministry for Research and Technology and the Federal Ministry for the Environment, Nature Conservation and Reactor Safety, in agreement with the departments concerned. Reappointment is possible.

(3)
Members may declare their resignation from the Council in writing at any time.

(4)
If a member resigns before the end of his or her term of office, a new member shall be appointed for the retired member's term of office.

Article 4

(1)

The Council is bound only to the brief defined by this Decree and is otherwise independent to determine its own activities.

(2)

Members of the Council may not be members either of the Government or a legislative body of the Federal Republic or of a Land or of the public service of the Federal Republic, of a Land or of any other juristic person under public law unless he or she is a university professor or a staff member of a scientific institute. Furthermore, they may not be representatives of an economic association or an employer's or employee's organisation, or be permanently attached to these through the performance of services and business acquisition. They must not have held any such position during the year preceding their appointment as member of the Council.

Article 5

(1)

The Council shall elect a Chairperson and a Vice-Chairperson from its midst for a term of four years by secret ballot.

(2)

The Council shall set up its own rules of procedure. These must be approved by the two ministries in charge.

(3)

If there is a differing minority with regard to individual topics of the report then this minority opinion can be expressed in the report.

Article 6

In the execution of its work the Council shall be supported by a Secretariat which shall initially be located at the Alfred Wegener Institut (AWI) in Bremerhaven.

Article 7

Members of the Council as well as the staff of the Secretariat are bound to secrecy with regard to meeting and conference papers considered confidential by the Council. This obligation to secrecy is also valid with regard to information given to the Council and considered confidential.

Article 8

(1)

Members of the Council shall receive all-inclusive compensation as well as reimbursement of their travel expenses. The amount of compensation shall be fixed by the two ministries in charge in agreement with the Federal Ministry of Finance.

(2)

The costs of the Council and its Secretariat shall be shared equally by the two ministries in charge.

Dr. Heinz Riesenhuber
Federal Minister for Research and Technology
Prof. Klaus Töpfer
Federal Minister for Environment, Nature Conservation and Reactor Safety
May 1992

— Appendix to the Council Mandate —

TASKS TO BE PERFORMED BY THE Advisory Council Pursuant to Article 2, para. 1

The tasks of the Council include:

(1)

Summarising and continuous reporting on current and acute problems in the field of global environmental change and its consequences, e.g. with regard to climate change, ozone depletion, tropical forests and fragile terrestrial ecosystems, aquatic ecosystems and the cryosphere, biological diversity and the socioeconomic consequences of global environmental change. Natural and anthropogenic causes (industrialisation, agriculture, overpopulation, urbanisation, etc.) should be considered, and special attention should be given to possible feedback effects (in order to avoid undesired reactions to measures taken).

(2)

Observation and evaluation of national and international research activities in the field of global environmental change (with special reference to monitoring programmes, the use and management of data, etc.).

(3)

Identification of deficiencies in research and coordination.

(4)

Recommendations regarding the avoidance and correction of maldevelopments.

In its reporting the Council should also consider ethical aspects of global environmental change.

Index

Springer
and the
environment

At Springer we firmly believe that an international science publisher has a special obligation to the environment, and our corporate policies consistently reflect this conviction.

We also expect our business partners – paper mills, printers, packaging manufacturers, etc. – to commit themselves to using materials and production processes that do not harm the environment. The paper in this book is made from low- or no-chlorine pulp and is acid free, in conformance with international standards for paper permanency.

 Springer